D0461637

INFORMATION TECHNOLOGY
FOR CONSTRUCTION MANAGERS,
ARCHITECTS AND ENGINEERS

INFORMATION TECHNOLOGY FOR CONSTRUCTION MANAGERS, ARCHITECTS AND ENGINEERS

Trefor Williams
Rutgers University

THOMSON

DELMAR LEARNING

Australia Canada Mexico Singapore Spain United Kingdom United States

Information Technology for Construction Managers, Architects and Engineers

Trefor Williams

Vice President, Technology Professional Business Unit:
Gregory L. Clayton

Product Development Manager:
Ed Francis

Director of Marketing:
Beth A. Lutz

Production Director:
Patty Stephan

Marketing Specialist
Marissa Maiella

Marketing Coordinator:
Jennifer Stall

Production Manager:
Andrew Crouth

Development
Sarah Boone

Content Project Manager:
Kara A. DiCaterino

COPYRIGHT © 2007 Thomson Delmar Learning, a division of Thomson Learning Inc. All rights reserved. The Thomson Learning Inc. Logo is a registered trademark used herein under license.

Printed in the United States of America
1 2 3 4 5 RRD 08 07 06

For more information contact Thomson Delmar Learning
Executive Woods
5 Maxwell Drive, PO Box 8007, Clifton Park, NY 12065-8007

Or find us on the World Wide Web at www.delmarlearning.com

Library of Congress Cataloging-in-Publication Data

Williams, Trefor.
  Information technology for construction managers, architects, and engineers/Trefor Williams.
      p. cm.
  Includes index.
  ISBN 1-4180-3958-6
1. Building—Data processing.
2. Construction industry—Data processing. 3. Architecture—Data processing. 4. Information technology.
I. Title.
  TH215.W55 2006
  690.0285—dc22
            2006020367

NOTICE TO THE READER

TO NANCY

BRIEF CONTENTS

TABLE OF CONTENTS

INTRODUCTION

This book has been written as a tool for practicing construction managers, engineers, and architects who want to know more about the application of information technology in the construction industry. The scope of the book is broad, covering a wide variety of IT issues including computer software applications, 4D CAD, the Internet, knowledge management, computer networking, and automation. The book includes consideration of well-established IT solutions such as scheduling and estimating software along with emerging trends such as web portals, knowledge management, and mobile computing.

Most people in the construction industry are very busy. This book is intended to be a reference about possible IT applications in construction. This book has also been written to provide students in construction management classes at both the graduate and undergraduate level with an introduction to the use of computers in construction and to suggest emerging research areas.

REASONS FOR DEVELOPMENT

The approach of this book is to provide a comprehensive treatment of IT applications in the construction industry. Few books exist that address construction IT in its entirety. Additionally, the book covers areas that have emerged recently including web-based systems and the widespread use of wireless and mobile computing. This book aims to provide a reference for the full range of IT activities in construction.

Some of the IT systems considered in this book are very expensive and their use is probably limited to large construction projects. Another reason for the development of this book is to suggest inexpensive and easy-to-use applications for smaller projects along with the more complicated solutions. Discussions of technologies such as weblogs, content management systems, and peer-to-peer networking are included in the book to show the relative ease with which the new IT technology is applicable to all types of users.

Emerging Trends

The use of IT in the construction industry is developing at a rapid pace. This book includes emerging areas of IT application in the construction industry that have the potential to revolutionize how construction projects are managed, including information about 4D and 5D CAD.

This text responds to emerging trends in the construction industry by providing:

- Extensive coverage of the use of the Internet and web-based systems.
- A discussion of knowledge management and collaboration.
- Consideration of data exchange and integration of software.

Level of Expertise

A reader does not need extensive experience with computers to understand the concepts discussed in this book. The primary purpose of the book is to provide managers and students with an idea of what can be accomplished using IT. Anyone who has used a personal computer and a web browser can understand the concepts and examples presented in the book.

GENERAL ORGANIZATION OF THE BOOK

The book is organized in chapters that focus on specific IT topics of concern to construction managers. Chapter 1 provides an introduction and focuses on describing the basics of the Internet. Chapter 2 discusses how the emerging area of knowledge management can be applied in construction. Chapters 3 and 4 focus on two issues that have traditionally been computerized in the construction industry: estimating and scheduling. These chapters also describe the latest capabilities of scheduling and estimating software. Chapters 5 through 8 concentrate on the emerging uses of the Internet to manage construction projects and exchange information and knowledge. These chapters also include discussion of web portals, weblogs, peer-to-peer networking, and online bidding. Chapter 9 describes how 3D, 4D, and 5D CAD are employed in the construction industry. Chapter 10 discusses how accounting software is employed in the construction industry. Chapters 11 and 12 deal with the emerging issues of mobile computing, wireless connectivity at the construction site, the automation of construction equipment, and the application of sensors at the construction site. Finally, Chapter 13 provides a roadmap for the implementation of IT based on a firm's familiarity with IT. Most chapters are structured to provide a general introduction to the IT topic, a discussion of existing applications, a discussion of some of the popular software or equipment available, and then some illustrated examples of the functions of software or equipment.

ACKNOWLEDGEMENTS

Many people have assisted me in the writing of this book. Valerie Watson and Joe Phelan of InfoTech provided extensive demonstrations and information about the line of InfoTech programs and the Bid Express Online bidding system. Nicholas Johnson and Tracy Murphy of Constructware provided access to the Constructware web portal and good examples of the uses of Constructware. Dominic Gallello of Graphisoft took the time to answer my questions about 4D and 5D CAD and provided examples of 4D CAD applications. Greg Duyka, Heather Charlesworth and Crystal Barger of On Center software helped me to learn about paperless quantity takeoff. Amy Urban of Tripod Data Systems helped me find a good picture of a ruggedized PDA. John Heinz of ATSG helped me understand the technical details of Blackberry PDAs and provided screenshots of a construction application using the Blackberry. Malcolm Davies of Gehry Technologies provided pictures and information about 4D CAD. Kelly Henry of Primavera answered several questions I had about the various Primavera software offerings and helped me find good pictures of the capabilities of the Primavera scheduling software. Chris Connolly and Hans Josef Kloubert of BOMAG provided information about BOMAG's automated compaction systems. Dean Bowman and John Schown of Bentley Systems provided valuable information about Bentley's construction related products. Daniel Wallace of Trimble helped me understand how GPS can be used to monitor construction equipment. Christian Grill of Engius provided a picture and case study of the intelliRock system. Pete Quintas of SilkRoad provided information about SilkBlogs and Truelook. Pete also graciously donated the use of the SilkBlog web service to allow us to experiment with the use of weblogs on construction projects. Jay Shapiro allowed us to use weblogs on one of his projects, and provided example schedules for this book. The help of Arne Aakre and Jennifer Sokoloski of Jay Shapiro and Associates in implementing the weblog is gratefully acknowledged. Doug Couto of the State of Michigan provided many interesting discussions about IT, information about Field Manager and many helpful comments about planning IT implementations. John LaPadula of Lockwood-Greene helped me understand how scheduling is used in the construction industry.

I would like to thank several people for their encouragement during the writing of this book. My friend Eleanor Fried always provided me with good advice. My friend and colleague Ali Maher, the Chairman of the Civil Engineering Department at Rutgers University, encouraged and advised me throughout the writing of this book. My wife, Nancy, was, as always, a great help and inspiration.

Thomson Delmar Learning and the author would also like to thank the following reviewers who provided valuable comments during the development of the manuscript:

Boong-yeol Ryoo, Florida International University
Chul Kim, Indiana University-Purdue University Indianapolis
Douglass Couto, Michigan Department of Information
James Jenkins, Purdue University
Joe Phelan, Info Tech., Inc.
Kirk Pickerel, Associated Builders & Contractors, Inc.
Mike Dunbar, Associated Builders & Contractors
William Whitbeck, Michigan State University

ABOUT THE AUTHOR

Trefor Williams is a Professor of Civil Engineering at Rutgers University. He has taught construction management at Rutgers for 19 years. He received his PhD in Civil Engineering from the Georgia Institute of Technology in 1987. He also holds an M.S. in Civil Engineering from Georgia Tech and a B.S. in Civil Engineering from Syracuse University. He has several years of experience as a project engineer for highway and traffic signal construction projects where he was responsible for construction quality control and implementation of traffic control systems. His research interests include mobile computing, web-based applications for construction management, dredging operations management, and analysis of construction project bidding using artificial intelligence and statistics. He is a member of the Transportation Research Board's Construction Management Committee, the ASCE Construction Research Council and the ASCE Construction Institute's Wireless Construction Committee. Dr. Williams is a Professional Engineer in New York and New Jersey.

CHAPTER 1

Computers and the Construction Industry

INTRODUCTION

Computers and communications technology have the potential to transform the construction industry by providing improved information to managers in the field. Wireless communications and networking technologies now make it possible to provide instant collaboration between project team members. This chapter will provide an introduction to the Internet, computer networking, and computer applications in construction.

■ Computers and the Construction Industry

New developments in computing are making powerful computing technologies available at low cost. There is the potential to provide advanced Information Technology (IT) techniques to areas of the construction industry that have never previously enjoyed the productivity benefits of computer use. In particular, computing solutions are becoming available for small construction contractors that were previously used only by the very largest companies. Computers have long been used in the construction industry, particularly in the areas of scheduling and estimating. The continuing development of computing power, the emergence of the Internet, and developments in wireless and mobile computing provide exciting opportunities to expand the use of computers and IT in the construction industry (See Table 1-1 for a list of advantages and concerns of IT use in the construction industry).

TABLE 1-1 Advantages of IT Implementation and Adoption Issues In the Construction Industry

Emerging Uses of IT in the Construction Industry	IT Adoption Issues in the Construction Industry
• It is accepted that the increasing application of IT can lead to productivity gains in the United States construction industry • Uses are evolving rapidly that can increase construction productivity • Web-based systems are promoting the exchange of construction information and knowledge • The web allows smaller companies to implement IT solutions • Mobile computing and wireless networking can bring computers to the construction project	• Industry has been slow to adapt many useful computer technologies • Lack of knowledge by people in the construction industry about what is possible • Firms exist along a spectrum from highly computer literate to those that hardly use computers • Industry is highly fragmented, making solutions that work on large projects infeasible for small projects • Sophisticated technologies and web services like web portals have been too expensive for small projects and small contractors

■ Definition of IT and Web-Based Systems

Information Technology (IT) can be defined as the acquisition, processing, storage, and dissemination of all types of information using computer technology and telecommunication systems. The initial applications of IT in construction focused on the use of computers for Critical Path Method (CPM) scheduling and estimating. Recently, web-based systems such as web portals for managing project information have emerged. The development of IT is continuing and provides many exciting opportunities to improve project management by providing enhanced information exchange and collaboration.

The development of the Internet has greatly increased the flexibility of how computer services can be supplied to users. In the past, any program that was used had to be provided on each user's computer, making collaboration between users difficult. With the emergence of web-based systems, IT solutions can be delivered via the web, which means that powerful computing capabilities often can be delivered to users with only a web browser. Web-based systems allow IT solutions to be easily accessible and easily distributed over the web.

Web-based systems can be efficiently administered with any modifications to the system instantly accessible to all users. Another consideration with older systems is that all users' hardware had to be compatible with the system. Providing IT solutions via the web allows users with different types of computers to easily access the information and knowledge.

■ The Internet and the World Wide Web

The use of the Internet and World Wide Web (WWW) will figure prominently in many of the discussions in this book. Most readers of this book have probably used a web browser to get information from the WWW. However, it is appropriate to provide a clear definition of the Internet and the WWW; an explanation is provided in Table 1-2.

TABLE 1-2   Internet and the World Wide Web

Internet	World Wide Web (WWW)
• Collection of public computer networks that are interconnected across the world by telecommunications links • Comprised of millions of computers world wide including businesses, government agencies, universities, and personal users • Allows users to communicate with each other using software interfaces such as electronic mail, File Transfer Protocol (FTP), Telnet, and the World Wide Web	• Part of the Internet where connections are established between computers containing hypertext and hypermedia materials • A WWW browser provides universal access to the materials that are available on the WWW and the Internet

Basics of the Internet and Client/Server Computing

There are several computing areas in which construction managers must have a basic understanding if they are to make informed decisions about the computer equipment and software they acquire. Increasingly, decisions about the types of computer applications that can be employed on construction projects must include knowledge of computer networking. Construction managers must know about client-server networks, and the basic functioning of the Internet. Networks are widely used in the construction industry from a small office network, to large networks that tie together the geographically dispersed offices of an international contractor. Many of the most popular construction software programs can now be networked in a Local Area Network (LAN) and/or accessed through the Internet, and construction managers must know the capabilities of these technologies. A basic definition of client/server computing is that it divides a computer application into three components: a server computer, client computers, and a network that connects the server computer to the clients (Lowe 1999).

Clients and Servers

One definition for a server is a computer that delivers information and software to other computers linked by a network. Another definition for a server is a computer that handles requests for data, e-mail, file transfers, and other network services from other computers (i.e., clients). Personal computers are typically used as clients. The types of computer equipment used for servers are more varied. A server can be any computer from a PC to a very large array of computers or a mainframe computer. The selection of the server will vary depending on the application and the number of users.

Types of Networks

There are several types of data networks that are employed to connect computers. They are:

- Local Area Networks (LANs)—A group of computers, and other computer devices such as printers and scanners that are connected within a limited geographic area. LANs are the type of network used to connect computers together in an office or project trailer (Dodd 2005). Figure 1-1 shows the basics of a LAN. LANs are configured to allow users access to the local network and the Internet.

Computers and the Construction Industry

- Wide Area Networks (WANs)—A group of LANs that communicate with each other. WANs are employed for connecting LANs that are widely separated geographically. WANs are commonly employed by large construction companies to connect their offices together. The LANs that compose the WAN can be connected together in a variety of ways including connection to the Internet and private high-speed data connections. A basic WAN configuration is shown in Figure 1-2.
- Metropolitan Area Networks (MANs)—Networks that can communicate with each other within a city. We will return to MANs when we discuss WiMAX wireless networking in Chapter 10.

To implement client/server computing using a LAN or WAN requires network operating system software to be installed on the server. The client computers must also have compatible operating system software installed. There are many well-known providers of server software, including Novell, Sun, Microsoft, and Hewlett-Packard. There is also open-source server software that is widely used. The type of software used for the server software can be dictated by the requirements of the application software you want to run on your network.

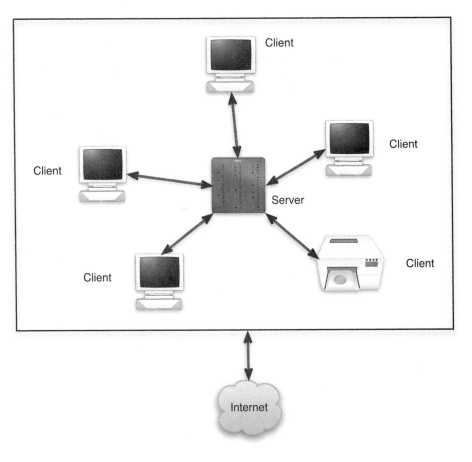

FIGURE 1-1 Typical Local Area Network

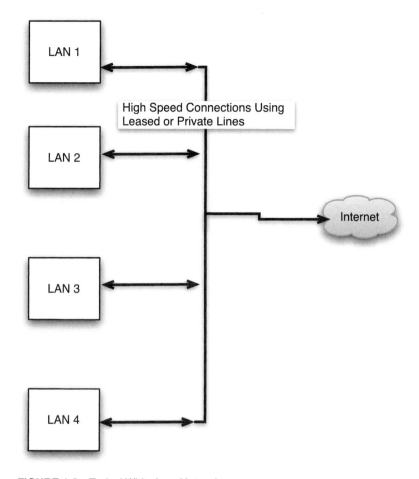

FIGURE 1-2 Typical Wide Area Network

More About the Internet and the WWW

As we discussed in the previous section, the Internet is the publicly accessible worldwide system of interconnected computer networks. It is the largest of all WANs and is made up of thousands of smaller commercial, academic, domestic, and government web sites (Wikipedia 2005). The purpose of the WWW is to make using and navigating the Internet easier. It is a graphical way of navigating the Internet using a web browser such as Microsoft Internet Explorer or Firefox. The WWW allows any type of computer with a browser to access information without needing to type computer commands (Dodd 2005). The ease of use of a web browser has fueled the expansion of Internet applications. The Internet is based on a client-server model in which a client with a web browser accesses web pages, e-mails, and files from a server. A set of protocols called Transmission Control Protocol/ Internet Protocol (TCP/IP) governs how the Internet operates. These include familiar protocols like FTP for file transfer and hypertext transfer protocol (HTTP) that is used to connect to web sites.

Intranets

Intranets are a computer network based on Internet technology that is contained within an enterprise. An intranet is implemented on a LAN. It can consist of many interlinked local area networks and also use leased lines in the WAN. It may or may not include connections through one or more gateways to the outside Internet. The main purpose of an intranet is to share company information and computing resources among employees. The intranet is accessible only by the organization's members, employees, or others with authorization. An intranet's web sites look and act like any other web site, but access to an intranet is restricted to the company's users and a firewall fends off unauthorized access (Precise Cyber Forensics 2005). An intranet allows users to work in the familiar web browser. We will discuss several applications that would work well on a construction company intranet. These include the web-based applications for information and knowledge management discussed in Chapters 5 and 7. Web portals are discussed in Chapter 6.

■ The Benefits of IT Use

It is generally accepted that the increasing application of IT can lead to productivity gains in the United States construction industry. However, the construction industry has been slow to adapt many useful computer technologies. Some of the reasons include:

- Lack of knowledge by people in the construction industry about what is possible. People in the construction industry are typically busy managing projects and do not have the time to keep abreast of the latest technology.
- Construction firms exist along a spectrum from highly computer literate to firms that hardly use computers at all.
- The construction industry is also highly fragmented, making solutions that work on large projects infeasible for small projects.
- Large contractors often have the financial resources to invest in sophisticated services like web portals that are too complicated or expensive for small contractors.

It is often difficult to determine the benefits of IT implementations. However several academic studies have confirmed that IT has several positive benefits on the outcomes of construction projects. Thomas and colleagues (2004) have performed an extensive study of the quantitative benefits of IT investment. Surveys were sent to owners and contractors regarding their use of design/information technology. The results were measured by determining the extent to which firms applied four technologies: integrated database, electronic data interchange, three-dimensional computer-aided design (3D CAD), and bar coding. The extent that firms employed these technologies were compared to five important project outcomes: cost, schedule, safety, changes, and field rework. Data from 297 United States projects were analyzed. The study found that project size is the single most important factor in determining the degree of IT use on a project. Large projects tended to have higher levels of IT use. The study also found that both owners and contractors obtain significant benefits from the use of IT. Finally, the study found that there are pronounced learning curve effects noticeable on many projects as project team members experiment and learn new technologies.

O'Conner and Yang (2004) studied data from 210 construction projects. They also studied the correlation between project outcomes and found that high levels of IT applied to high-tech work functions on construction projects is associated with high levels of cost success on projects. That is, projects that made extensive use of IT technologies tended to have less chance of having high cost overruns than projects that made limited use of IT.

These academic studies indicate there are benefits to using IT. Additionally, studies find that large projects tend to have a higher level of IT usage. Perhaps this indicates that methods should be found to allow smaller projects to reap the benefits possible from the application of IT.

How is IT Currently Used in Construction?

Hua (2005) reported on a survey of the application of IT to the construction industry in Singapore and compared these results to data on IT use by the construction industry in Scandinavia. What is interesting in this study is the reported statistics of web use. The Singapore data were collected in 2003 and the Scandinavian data in 2000 and 2001. The data found is listed below.

- Most construction companies have access to the Internet.
- 67% of Danish companies and 45.2% Singaporean companies reported maintaining their own company web page.
- A much smaller percentage of companies use web portals for storage and transfer of project documents.
- In Singapore, 35.7% of survey respondents reported using a web portal in 2004, whereas 25% of Swedish companies reported using a web portal in 2000.

Clearly, large segments of the construction industry do not yet use the web or its collaborative capabilities. Large construction companies are the first adaptors of new IT technologies because they have the resources to implement complicated IT solutions and because administration of the complex projects they undertake is greatly aided by IT adoption. The results of the IT surveys indicate that there are many construction companies that can benefit from implementing IT but have not yet done so because of financial constraints, or the perceived difficulties of understanding new technology.

Lim (2001) has discussed measures of the informatization level of an organization. Lim has postulated that organizations exist in a broad range of capabilities to assimilate and use information technology: their level of informatization. Figure 1-3 shows the spectrum of informatization commonly seen in the construction industry. The level of informatization varies widely in the construction industry. The capabilities of construction contractors to use computers range from those with only the most rudimentary skills to firms with sophisticated web-based and mobile computing applications. This variety in the capabilities of contractors indicates the need for easily implementable IT solutions that can be assimilated and used by a larger group of contractors.

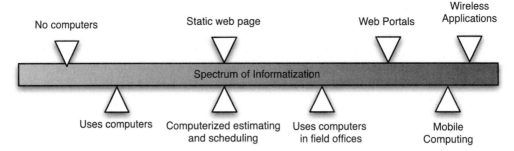

FIGURE 1-3   Evolution of Construction Firm Informatization

It seems that the continuing development of IT technologies will lower the barriers to IT use. Several sections of this book highlight easily implementable IT solutions that are useful for smaller companies and smaller-sized projects. Technologies that provide methods of computer-based collaboration and mobile computing are only beginning to be widely used.

Rivard and colleagues (2004) have studied the use of IT in the Canadian construction industry. They studied the use of computers on several major Canadian projects for design and construction. They found that construction web portals, scheduling software and 2D/3D CAD are widely used. They also found that some companies have developed their own customized web sites to serve as repositories for their internal documents. However, they found that none of the projects employed wireless or mobile computing. Additionally, the study found that none of the projects took advantage of the benefits of interoperability between software using standardized data formats. The study also noted that differences in IT capabilities between the different partners on a construction project could often cause difficulties. Additionally, many subcontractors may have minimal IT capabilities and full use of an IT system requires more technically adept project team members to enter the subcontractors' data.

■ Promoting IT: Adoption Within the Construction Firm

Clearly, some construction companies are more adept at implementing new IT technologies. Therefore, it is important to consider the forces that drive construction companies to adopt new technologies. There are four main forces that drive the adoption of new technology by construction companies (Mitropoulos and Tatum 2004). These forces are:

- the desire to seek competitive advantage
- the need to solve construction process problems
- external requirements to implement new technologies
- technological opportunity

Owners mandate the use of many IT solutions. The requirement by owners to implement web portals has caused many contractors to implement this technology. With the continuing development of web-based systems, new technological opportunities are continually appearing that will allow new IT systems to spread to construction companies of all types and sizes. A contractor adopting a new IT technology before its peers, may provide a productivity advantage, allowing the company to gain more business.

Mitropoulos and Tatum (2004) lists factors that allow for new technologies to be fostered in a construction company. Two important factors they discuss are the attitude of company management to adopting new technologies, and the allocation of resources within the construction company to implement the new IT solution. It has been discussed by Whyte and colleagues (2002) that the successful implementation of IT requires both strategic decision making by top management and decision making by technical managers. First, top management must make the strategic decision to purchase and use the new technology. Second, the technical managers within the construction firm (project managers and engineers) must determine how to incorporate the new technology into the firm's normal project workflow. The cooperation of technical managers is essential to successful implementation. Upper management must provide technical management of the resources (equipment, training, etc.) to allow for successful implementation.

The adoption of new IT technologies tends to succeed at companies where the top management understands the importance of IT to construction and at companies that have a culture that fosters innovation. Finally, it is important to provide the proper resources, including money, manpower, and training to successfully implement new technologies into a firm's practice. Often, the use of new IT technologies fails because personnel are too busy with other tasks or have not been given the proper training to fully reap the benefits of the new system.

Peansupap and Walker (2005) conducted a literature survey of academic papers about IT applications in construction. This survey listed five main areas that must be addressed to promote the successful implementation of construction IT projects. These factors are:

- Self-motivation. Users of IT systems must be able to perceive the benefits of the IT implementation.
- Training. Users must be provided with the appropriate training and technical support to use the new IT system.
- Technology Characteristics. New technologies that can be easily integrated into existing systems and that are not highly disruptive to user's workflow have the best chance of success. In other words, a new IT system that offers technology that is radically different from a firm's existing systems may be difficult to implement.
- Workspace Environment. The management of the construction company must be committed to the IT implementation and enough resources must be provided for the system to be implemented.
- Sharing Environment. An environment in which learning and sharing among the staff is believed to be important in implementing new IT systems.

Adequate training and a positive attitude toward the new IT system are important factors in determining the successful implementation of IT in a construction company.

■ The Universe of Construction IT Applications

When seeking to implement IT within a construction company, one must consider three major questions:

- What types of IT applications have historically been used in the construction industry?
- What are the newer types of applications that can now be used in the construction industry?
- What computer techniques will evolve in the near future that will be useful in the construction industry?

Figure 1-4 shows the range of available IT uses in the construction industry. More traditional applications are shown on the left of the figure with newer technologies shown on the right. Initially, computers were used in the construction industry to schedule and plan projects, produce estimates and budgets, and perform accounting for construction companies. There are many advanced programs now available to perform these functions. These applications can be considered the traditional use of computers in construction.

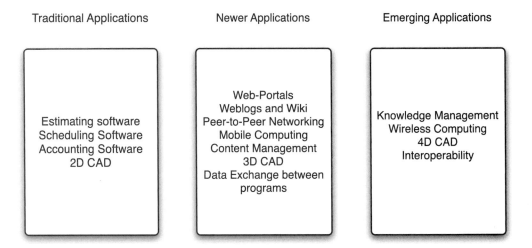

FIGURE 1-4 Advances of IT Implemetation and Adoption Issues in the Construction Industry

With the emergence of the Internet, applications that employ the power of the WWW have emerged. The benefits of the web include efficient distribution, effective administration, and cross-platform accessibility of computer software (Moelenaar and Songer 2001). In particular, web-based software allows access to complex computer software without the need to install the program on individual computers. Construction web portals have become popular. However, in this book we will discuss some simpler methods to use the power of the Internet, such as weblogs and peer-to-peer networking that may be more appropriate for smaller contractors.

Mobile computing is just beginning to be used in the construction industry. Some good examples of its use already exist. The next step, which is now emerging with technical advances in wireless networking, is the capability to provide everyone with wireless access to the Internet, even if they are working at a remote construction site.

The use of CAD is widespread in the construction industry. 3D CAD is already being used to check for conflicts and to provide walk-throughs of facilities before they are constructed. There are already commercially available 4D CAD products. 4D CAD shows a simulation in 3D of how a facility evolves over time. This is an IT capability that will continue to evolve and be more widely used in the future. Automation of construction equipment is another interesting development. Equipment manufacturers already use Geographic Information Systems (GIS) and computers to control grading and compaction equipment. This trend is sure to continue to evolve.

It is also possible to consider the typical functional areas within the construction company, and what types of IT applications can be used by them. Table 1-1 illustrates the various main functions within a construction company and the IT solutions that can be applied. The top management of a construction firm is interested in the firm's financial performance, and tracking ongoing progress to check cost and schedule status. Top management requires accounting software to produce the firm's financial statements and monitor the cost status of projects in progress. Top management must also communicate with the field and often use web portals or internal company content management systems.

TABLE 1-3 IT Solutions by Construction Firm Function

Management Function	Top Level Management	Estimating	Scheduling and Planning	Project Management	Operations Management
Typical Tasks	1. Company Financial Management 2. Information Exchange with projects and clients	1. Estimates of Project Costs 2. Bidding	1. Project Scheduling 2. Identification of conflicts	1. Cost Control and Schedule Conformence 2. Information Exchange 3. Field Data Collection	1. Control of Construction Operations 2. Specification Performance
IT Solutions	Accounting Software	Estimating Software	CPM Software	CPM Software	Knowledge Management
	Web Portals	Automated Quantity Takeoff	Monte Carlo Simulation	Accounting software	Electronic Books
		Interoperability with CAD Documents	3D CAD	Web Portals	Mobile Computers
			4D CAD	Mobile Computers Document Management Content Management Wireless Computing	Content Management Wireless Computing

Computers and the Construction Industry

As was discussed above, estimating software and CPM scheduling software has been a cornerstone of construction IT applications for many years. Most companies commit personnel full time to the estimating and scheduling functions. The software that is used has become increasingly sophisticated over time and new developments, such as paperless quantity takeoff, are already a commercial reality.

Additionally, it is now often possible to exchange information between many scheduling and estimating software programs, reducing the amount of time required for data input. The management of large projects has been greatly impacted by the development of web-based construction web portals that allow for communication, document management, and document exchange for construction projects. The potential for wireless computing to provide construction managers in the field with the latest project data is beginning to become a reality. So far the focus of construction IT computer systems has been on the exchange of project documentation.

Little work has been done on providing IT solutions that help construction managers organize construction operations or to provide knowledge about how to improve construction quality. Various IT knowledge management techniques show great promise to capture the knowledge that a construction company has so they can be reused in the field to aid less experienced managers.

Planning the Implementation of IT Projects

For a large construction company or designer, implementing many of the IT topics discussed in this book can be a complex process. There are reported instances in many industries where the implementation of IT has been unsuccessful or did not meet the expectations of end users. Successful IT projects require considerable planning. Many IT implementation projects will require the management of construction and design firms to work closely with IT experts. To provide systems useable by construction managers in the field, it is necessary that the activities of the IT staff are integrated with the needs and plans of the construction company. The plans for new IT implementations must be integrated with the strategic plans of the company.

Parker and colleagues (1988) have defined the Enterprise-wide Information Management (EwIM) method for planning the implementation of IT within a firm. The EwIM process includes four planning activities:

- Alignment. The plans for new IT systems must be aligned with the needs of the business organization. Before implementing new IT systems, it must be determined that they address existing deficiencies.
- Opportunity. As a business uses IT, new opportunities for the use of IT can be identified, and these needs can be integrated with the company's strategic plans.
- Organization. Planning must take place to determine how IT implementation efforts should be organized and identify the participants. In a construction context, it must be determined what IT, managerial, and field personnel must be involved to implement new IT systems.

- Impact. The implementation of new IT technologies may impact a firm by opening up new business opportunities and new customers. This impact will lead to the need for further consideration of IT needs.

These planning relationships are circular. In a large, dynamic construction organization these planning areas must be continuously addressed. New technologies will influence business strategies and plans. Good communications between a firm's IT experts and its management are essential to the success of this planning process and to insure that IT implementations are successful.

■ Conclusions

Various studies of IT use in the construction industry have shown:

- High levels of IT use among construction industry participants correlate with positive project outcomes.
- IT use is greater on large projects than smaller projects.
- IT can be used to increase a construction firm's productivity and further study is needed of methods of expanding the use of IT to all participants in the construction process.

The available tools for IT include traditional applications such as scheduling and estimating software. With the development of the Internet, new web-based applications are emerging that can provide new ways of exchanging information and knowledge. Finally, advanced technology is emerging that will allow for 4D scheduling and greater capabilities for data exchange between construction computer applications. These topics will be explored in the following chapters.

■ References

Dodd, Annabel. 2005. *The Essential Guide to Telecommunications*. Upper Saddle River, NJ: Pearson Education, Inc.

Hua, Goh Bee. 2005. IT Barometer 2003: Survey of the Singapore Construction and a Comparison of Results. *Electronic Journal of Information Technology in Construction* 10, http://www.itcon.org/cgi-bin/works/Show?2005_1 (accessed September 17, 2005).

Lim, Soo Kyoung. 2001. A Framework to Evaluate the Informatization Level. In *Information Technology Evaluation Methods and Management*, ed. Wim Van Grembergen, 144–153. Hershey, PA: Idea Group Publishing.

Lowe, Doug. 1999. *Client/Server Computing for Dummies*. Foster City, CA: IDG Books Worldwide, Inc.

Mitropoulous, P. and Tatum, C.B. 1998. Forces Driving Adoption of New Information Technologies. *Journal of Construction Engineering and Management* 126:5, 340–348.

Molenaar, K. and Songer, A. 2001. Web-Based Decision Support Systems: Case Study in Project Delivery. *Journal of Computing in Civil Engineering* 15:4, 259–267.

O'Conner, J.T. and Li-Ren Yang. 2004. Project Performance versus Use of Technologies at Project and Phase Levels. *Journal of Construction Engineering and Management* 130:3, 322–329.

Parker, Marilyn, M., Benson, Robert J., and Trainor, H.E. 1988. *Information Economics: Linking Business Performance to Information Technology*. Englewood Cliffs, NJ: Prentice-Hall, Inc.

Peansupap, V. and Walker, D.H.T. 2005 Factors Enabling Information and Communication Technology Diffusion and Actual Implementation in Construction Organizations. *Electronic Journal of Information Technology in Construction* 10, http://www.itcon.org/cgi-bin/works/Show?2005_14. (accessed September 19, 2005)

Precise Cyber Forensics. 2005. Glossary for Computer Forensics. http://precisecyberforensics.com/glossary.html (Accessed November 12, 2005).

Rivard, H., Froese, T., Waugh, L.M., El-Diraby, T. et al. 2004. Case Studies on the Use of Information Technology in the Canadian Construction Industry. *Electronic Journal of Information Technology in Construction* 9, http://www.itcon.org/cgi-bin/works/Show?2004_2 (accessed September 17, 2005)

Thomas, S.R., Lee, S.H., Tucker, R.L., and Chapman, R.E. 2004. Impacts of Design/Information Technology on Project Outcomes. *Journal of Construction Engineering and Management* 130:4, 586–597.

Whyte, J., D. Bouchlaghem and T. Thorpe. 2002 IT Implementation in the Construction Organization. *Journal of Engineering, Construction and Architectural Management* 9: 5/6, 37.

Wikipedia. 2005. Internet. http://en.wikipedia.org/wiki/Internet (Accessed November 9, 2005).

CHAPTER **2**

Knowledge and Information Management for Construction

INTRODUCTION

The emergence of new computer tools is opening up new ways to apply computers in construction. To date, construction applications have focused on handling information and data. New web-based computer techniques are now making it possible to consider using IT as a method of capturing and sharing knowledge about construction. This chapter focuses on the emerging area of knowledge management and how it can be applied in the construction industry.

◼ Knowledge and Information

Several important questions must be considered when we discuss information management, knowledge management, and IT's application in the construction industry. First we must consider the basic definition of knowledge and how it relates to information. We must also consider that the focus of existing applications in the construction industry has focused on managing information, whereas the idea of bringing knowledge management to construction involves a major shift in thinking. It is important to consider the difference between information and knowledge. Information is an assembly of data that is useful for a particular analysis or decision task (Norman 1998). Knowledge can be defined as information in context with an understanding of how to use it (Brooking 1999). Knowledge management can be defined as the processes in which knowledge is created, acquired, communicated, shared, applied, and effectively utilized (Egbu and Botterill 2002).

As we have discussed, IT is being increasingly used in the construction industry. Some of the issues and uses of IT in the industry are listed in the following section:

- Complex IT systems and web portals have been employed for the management of large projects. Web portals have improved the capability of all parties in the construction process to organize project information that includes the many thousands of documents that can be generated during a large and complex project
- Existing computer programs and web-based applications have focused on the flow of information and data, but have not been explicitly concerned with the management of knowledge. Interest is increasing in methods of retaining and disseminating a construction company's knowledge.
- The emergence of complex estimating and scheduling systems that can organize and prepare the required information to plan a project or prepare a bid estimate. The trend is to develop integrated computer applications that can share information and data.

The industry lacks computer systems that help construction managers interpret the large amounts of information available. IT has so far lent greater organization to our information. We must now consider how to use IT to provide construction managers with knowledge on how to best utilize and interpret the available information and manage projects effectively.

The organization of information is vital to the management of construction companies and projects. However, it has been increasingly recognized that companies must also manage their knowledge to remain competitive. With many firms employing similar IT systems for the management of information, using IT for knowledge management can potentially allow a firm to differentiate itself from its competitors by having a "smarter" organization that is able to quickly access its best practices and knowledge whenever issues arise throughout the construction process.

■ Introduction to Knowledge Management

There has been considerable recent interest in the application of knowledge management to business organizations. Researchers in knowledge management have recognized that knowledge may be a company's most important competitive asset. Companies now require quality, value, service, innovation, and speed to market for business success. Increasingly, companies can differentiate themselves by what they know (Davenport and Prusak 2000). This realization has led to considerable discussion among academics and practitioners about how the knowledge assets of a firm are best employed (Egbu and Botterill 2002). The construction industry is a fertile area for the application of knowledge management. Clearly construction companies can be more competitive by better managing their knowledge and communicating it effectively to personnel in the field.

It is widely recognized that the construction industry is characterized by the existence of subject-matter experts. Knowledge can be lost when an expert leaves a company. Because construction is performed on a project basis, it is also possible to lose knowledge from a project that could be gainfully employed on another project. Within small construction companies, very little documented material may exist about the firm's construction practices. Instead, most knowledge is contained in the minds of a small group of the firm's principals.

Alshawi and Ingirge (2003) have noted that there is a lack of standard processes for project management. Each project manager in an organization prefers to follow her experience. Potentially, the application of knowledge management principles can provide construction companies with the capability to share more knowledge about the best procedures for conducting construction operations and to standardize the best procedures. This could result in improved productivity and quality. Perhaps, the more widespread application of knowledge management could foster increased innovation.

In Japan, it has been recognized that creating new company knowledge can be facilitated by tapping the tacit insights, intuitions, and hunches of employees and making those insights available for testing and use by the company as a whole (Nonaka 1998). Perhaps, a more pervasive use in the construction industry can provide a catalyst for innovation and new ways of conducting construction.

IT is defined as the "acquisition, processing, storage, and dissemination of all types of information using computer technology and telecommunication systems (ASLIB 2003)." Knowledge is defined by Davenport and Prusak (2000) as a "...mix of experience, values, contextual information, and expert insight that provides a framework for evaluating and incorporating new experiences and information." In organizations, knowledge is not only embedded in documents but also in organizational practices and processes. IT is often seen as the means of implementing and applying knowledge management within companies. Tiwana (2002) describes the need for the smart distribution of knowledge so employees can find existing critical knowledge in time and share lessons learned. Clearly, the construction industry can benefit from providing better knowledge to employees.

◼ Examples of Knowledge Management from Other Industries

Many industries have begun to employ knowledge management to more effectively use their assets. Jenkins (2004) provides several examples of computer systems used by large companies to manage knowledge. BT, the British telecommunications company, provides a project workspace system to capture and reuse its intellectual assets and to enable the establishment of e-communities to increase collaboration. The increased communication and exchange has facilitated common understanding amongst BT employees. The pharmaceutical company Roche uses a system called ShareWeb to support global team projects.

Knowledge and documents from previous projects are available and updated regularly. The system has been found to allow new project teams to be assembled faster. It is often mentioned in the knowledge management literature that information tends to be stored in "silos" within organizations. This means that organizations tend to become compartmentalized, and expert knowledge does not get shared with other people in the firm. Air Liquide, a global company that specializes in industrial and medical gases has implemented a web-based knowledge management system for 12,000 users to provide a way of disseminating documentation and knowledge throughout the firm, to integrate "silos" of information, and to standardize the use of best practices. These three examples illustrate the seriousness with which large business enterprises take knowledge management and the extensive IT systems they are employing to facilitate the better use of knowledge within their companies.

■ Tacit and Explicit Knowledge

Different types of knowledge exist within the construction company. They can be defined as tacit and explicit knowledge. Typically, the knowledgeable people within the firm are promoted to higher management positions and may not be available in the field to solve problems. Table 2-1 lists the components of tacit and explicit knowledge.

TABLE 2-1   Tacit and Explicit Knowledge

Explicit Knowledge	Tacit Knowledge
• Relatively easy to capture and store in documents • Can be either structured or unstructured • Structured explicit knowledge is organized in a particular way for future retrieval • Examples would be documents and spreadsheets • Unstructured explicit knowledge includes e-mails and images that contain knowledge but are more difficult to retrieve (New York State Department of Civil Service 2002)	• Knowledge that people have in their minds, more difficult to access • Transfer of tacit knowledge involves personal contact and trust • Highly valuable, provides a context for ideas and experiences (New York State Department of Civil Service 2002) • Considerable amounts of tacit knowledge exist within construction companies • The tacit knowledge of experts is the most valuable asset of a construction company

There are several areas in which tacit knowledge is important within the construction company and for which the utilization of IT techniques could benefit the company. The capture of tacit knowledge can benefit a construction company in several ways:

- The knowledge possessed by experienced managers can be shared with less-experienced managers for educational purposes.
- Problems and their solutions can be collected for application to future projects.
- The capture of the firm's expert knowledge can protect the company from the loss of expertise through attrition or retirement.
- Knowledge management IT systems can make the firm's best practices available to a broader range of personnel, enhancing the firm's overall decision-making capabilities.

Knowledge management can be applied to several areas of a construction company's operation, including those listed in the following section.

1. **Operations**—Construction companies typically have little documentation about the best ways to organize and perform construction operations. Questions about the appropriate crew and equipment to perform construction tasks are often asked by inexperienced engineers in the field. Better access to knowledge about the construction company's standard procedures could

reduce construction mistakes. Additionally, techniques for control of construction quality are probably learned through experience and become tacit knowledge.

2. **Management of the firm and project**—Tacit knowledge about how to manage construction projects. For example, what should be done if some activities begin to lag? An experienced manager knows how to rearrange resources and activities to minimize problems. Can this information be captured and transmitted to less experienced people within the construction company?

3. **Bidding and Scheduling**—Certainly much tacit knowledge exists within construction companies about the correct procedures for estimating, bidding, and project planning. It can be used to capture the firm's best practices that are embodied in the tacit knowledge of the firm's senior personnel. How to perform activities such as the best procedures for estimating, knowledge of the firm's market, and how to best plan and schedule a complex project are typically learned through years of experience. Any system that can tap the tacit knowledge of the firm's most experienced estimators and planners and make that knowledge available to more people in the firm has the potential to improve the company's performance.

■ IT Systems for Capturing Tacit and Explicit Knowledge

Different types of IT systems have been found to be most suitable depending on the application and the type of knowledge that must be captured. The capture of explicit knowledge using IT is typically easier. Various types of easy-to-use, web-based IT systems have emerged that can be used to contain written documentation and manuals containing explicit knowledge. Harder to capture is the tacit knowledge within the construction company. Several of the web-based solutions discussed in this book have the potential to allow collaboration between construction company employees. Several industrial applications have been developed that focus on collaboration between users as a way of extracting tacit knowledge.

BP Exploration has developed a knowledge management system that uses video conferencing, document scanning, and collaboration tools to allow experts who may be in remote locations around the globe to interact when a problem occurs. BP Exploration decided that the important knowledge in the organization was unstructured and only in the heads of its employees. Therefore the focus was on ways of promoting collaboration rather than building a document repository (Davenport et al. 1997). The continuing development of IT and lower costs for video conferencing equipment and web conferencing will eventually place collaboration solutions like this in the reach of most construction companies.

■ Potential Uses of Construction Knowledge Management and Its Relationship to Information Management

Several examples illustrate the usefulness of knowledge management IT applications and their relationship to the information management applications that are already widely used in construction. Estimating software is used to automate the calculation of bid estimates. However, key information like the appropriate mark-ups and overhead rates to employ are known only to experienced estimators. A knowledge management system could document important bidding information and provide a reference for less-experienced estimators within the firm. Similarly, construction managers may know when various tasks and construction operations must be performed from the output of a scheduling program, but they may have little idea about how to accomplish the task. Asphalt paving provides a good example. Quality control of asphalt paving can be complex, requiring considerable experience to recognize paving problems and to suggest possible solutions.

A knowledge management system can suggest potential solutions for adjustments to paving machine settings to maintain quality. In the author's experience, constructions of asphalt pavements have been delayed while inexperienced engineers attempt to contact a paving expert who was responsible for several projects. Often the delay was corrected by only a small adjustment in paving procedures. If knowledge management was used, engineers at the job site could have accessed the required information rapidly, improving project productivity.

Construction companies can consider maintaining parallel IT systems and software. One system is to manipulate, transfer, and store information and the other system is to capture and document knowledge for reuse. For example, web portals can be usefully employed to manage the documents and information generated on a construction project. In parallel with that system, an IT system, perhaps a content management system or a collaborative weblog can exist to capture and store the knowledge about management and construction procedures necessary to complete the project. In this book we will discuss IT applications in both domains, the domain of information management and the domain of knowledge management.

A Case Study from the Automotive Industry and How It Relates to Construction

Ford provides a compelling case for the use of knowledge management. The original Taurus automobile was a highly successful vehicle. When Ford management wished to study why the project had been so successful it was found that no documentation about the project existed, project participants had been dispersed throughout the company, and the engineering knowledge that made the project so successful had been lost forever. Because the knowledge was lost, Ford was unable to duplicate the successful Taurus program (Tiwana 2000).

continued

Ford's story relates strongly to the construction industry. Most construction companies form a project team. This team learns the "lessons" for the particular type of project being constructed. After the project is complete, the team separates and is scattered about the company's other projects. When the company again performs a similar project, a new team is formed that must again relearn the "lessons" for that type of construction. The knowledge of the original team is lost and is not available to the new project team.

Barriers to KM Use

Figure 2-1 illustrates the relationship between information management and some of the barriers to implementing knowledge management applications for personnel monitoring construction operations. We have been successful in developing IT tools for information exchange; however, few examples of knowledge management systems exist in the construction industry. Figure 2-1 shows how tools such as web portals have been successful in allowing for information exchange. However, such tools are not intended to be repositories of a firm's knowledge; they are instead tools to foster document exchange and tracking of project documents.

Barriers exist to developing knowledge management systems for construction uses. First, busy construction managers often see recording lessons learned as an extra work task. Proper incentives most be provided to encourage participation in knowledge management systems. Any knowledge management system must be easy to use and provide a good user interface for the input of information. Knowledge management systems work best in a business culture where sharing of knowledge is encouraged. Many people may not be willing to share their knowledge. Therefore, incentives must be provided within the firm to encourage participation in knowledge management programs. Participation in knowledge management programs can be increased by offering bonuses for good ideas or by linking participation in a knowledge management system to an employee's evaluations. Additionally, expert knowledge within the construction firm is often concentrated among the top management and busy experts. It is often difficult for subject matter experts to find the time to contribute "lessons learned."

Finally, a good starting point in developing a knowledge management system is to incorporate any existing documentation about procedures in the KM system as an electronic book or hypertext web page. Unfortunately, many construction companies today do not maintain information about standard practices that could form the basis for a computerized KM system.

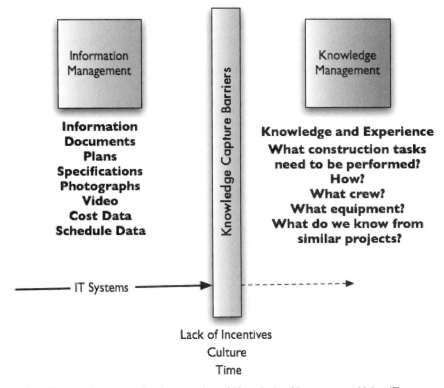

FIGURE 2-1   Barriers to Implementation of Knowledge Management Using IT

■ Some Suggested Areas and Techniques for Developing KM Systems

There are several good areas for developing KM techniques in construction. One possibility is to provide a knowledge base of the firm's best construction practices. Inexperienced managers are often confronted with questions about the best ways to conduct a construction operation. Many types of construction are complex and require expert knowledge to interpret what is being observed. For example, in asphalt paving, deviations in quality such as streaking or waves may be readily observed. However, the correct response may not be obvious to an inexperienced person. The construction process may be stopped until an experienced engineer can be located. Instead, a KM system available through the web or contained on a mobile device could provide knowledge about what to do to correct the problem. This can lead to improved productivity because there are fewer delays and improved quality because deviations from specifications are detected more rapidly.

Knowledge management also has a place in the home office of the construction company. Estimating is a vital concern for any construction company. Certainly a KM system describing the standard estimating practices of the firm and listing lessons learned from past projects would be useful in improving a firm's bidding practices.

▉ The Link Between Knowledge Management and Information Technology

A primary means of implementing knowledge management within a construction firm is computers and information technology. Many of the computer applications in this book can be readily used for knowledge management applications. Many different types of IT systems can be used to store the explicit knowledge available. Certainly, web portals already do a good job of this by providing a repository for project documentation. For tacit knowledge, weblogs, wiki, electronic books, and content management systems can all be used to capture a firm's knowledge and practices (These IT tools are discussed in detail in Chapters 5 and 7).

"Communities of Practice" as a Construction Knowledge Management Tool

One way to obtain tacit knowledge is by forming communities of practice among employees and providing access to subject matter experts. A community of practice can be defined as a learning network in which practitioners connect to solve problems, share ideas, set standards, and develop relationships (Snyder and de Souza Briggs 2004). The focus of a community of practice is to build and share knowledge among practitioners.

A community of practice is defined by three important elements:

- Purpose. The purpose of a community of practice is to cross traditional organizational and geographic boundaries to foster innovation in a specific topic area. A construction company could develop communities of practice for many topics, including estimating techniques or lessons learned from the field.
- Domain. A good community of practice would focus on a specific issue that can improve performance on construction projects. A good focus area for a construction company is some type of work that it performs frequently. Asphalt paving construction is a good example because paving can be complex and many different problems may occur during a construction project. The community of practice can serve as a way of unlocking the knowledge of paving experts.
- Community. Successful communities of practice allow for the active interchange of ideas and foster a feeling of commitment to other participants in the community. A primary use of a community of practice in construction is to allow less-experienced personnel to exchange ideas with domain experts.

Many of the emerging IT techniques discussed in this book are ideal ways to facilitate a community of practice by providing a platform to develop, share, and archive knowledge for use by communities of practice within a construction company, or for use by all of the participating organizations in a complex project. Web-based systems, such as weblogs and wiki, allow any authorized user to add comments and ideas to the system without the need to learn any computer programming skills. Additionally, the low cost and simplicity of implementing and using these web-based systems make it possible for construction companies of all sizes to use knowledge management techniques to their advantage. Perhaps the greatest challenge in the construction industry is to create an atmosphere in which employees are willing to freely exchange ideas and knowledge.

Using Concept Maps to Capture Construction Company Knowledge

A good example of the use of IT to support knowledge management is computer programs that allow concept maps of knowledge to be quickly generated. Concept maps are a graphical way of displaying and sharing knowledge. A concept map is a graphical two-dimensional display of concepts—the nodes in the diagram—linked by directed arcs. The arcs are encoded with a phrase that links the concept nodes together to form propositions (Canas et al. 2005). In other words, a concept map is a diagram illustrating knowledge in a particular domain through the links between various concepts. Concept mapping has been found to be an effective way of representing a person or group's understanding of a problem domain. Concept mapping may be particularly useful when applied to construction because it takes very little time to learn how to create concept maps, yet it can represent very complicated knowledge domains.

Concept maps have several possibilities for application in construction as a knowledge management tool:

- To capture expert knowledge about construction operations and processes. A concept map enables users to graphically express their understanding of construction processes.
- To brainstorm about the best way to address project issues and problems.
- To retain the knowledge of senior personnel by capturing it as a concept map.

The development of computer software has made it easy to create, manipulate, and use concept maps. One example of this type of software is the CmapTools program. CmapTools is a software program developed by the Institute for Human and Machine Cognition, a research institute of the Florida university system that is affiliated with several universities in Florida. The program and extensive documentation are available at http://cmap.ihmc.us/.

CmapTools allows for the development of concept maps using simple drag-and-drop tools. Links to all types of other resources can be established such as web pages, textual information, pictures, and video clips to enhance the information shown on the map.

A portion of a concept map to answer common questions about asphalt paving construction problems is shown in Figure 2-2. Notice the small icons that are attached to some of the concept nodes. These indicate that links to additional materials are available.

Figure 2-3 shows how if the link associated with the "Overcorrecting thickness controls" concept is selected, additional information is displayed about how to operate the paver controls.

The CmapTools program can work well as a stand-alone program on a single computer. However, it has several unique features that allow for collaboration and the ability to publish the concept map and its related materials on the web. There are two ways to publish the concept map to the web for use and collaboration:

- Public Places. Public Places are various computers running as Cmap servers. Any user on the Internet can create a folder on these servers that contains a concept map and its related files. The user who uploads the material becomes the administrator of the concept map and can control who has permission to use the concept map. There are different levels of permission from the ability to modify and delete map elements to viewers who are only given permission to make comments. Simultaneous collaboration between users is possible for concept maps on a Public Places server.

- Save the concept map as a web site. Alternatively, the CmapTools program can save the concept map as a web site in HTML format. The concept map can then be uploaded to a firm's private intranet or Internet web site.

The CmapTools program can link multiple concept maps for large problems. Perhaps a construction company could develop concept maps for all of its important operations and link them to provide access to the complete body of the firm's knowledge.

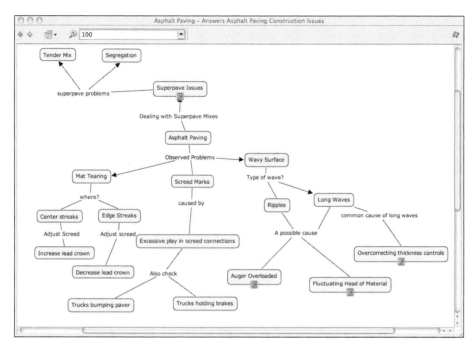

FIGURE 2-2   A Concept Map of Asphalt Paving Construction Issues

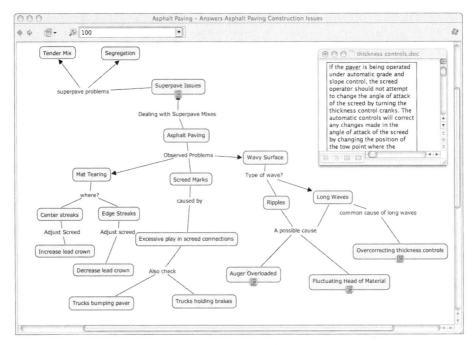

FIGURE 2-3 The Concept Map Displaying Linked Knowledge

■ Conclusions

It has long been recognized that computers are vitally important to organizing information in the construction industry. Recently, interest has focused on extending the use of computers beyond traditional applications to manage the knowledge of the construction. Many examples exist from other industries of how knowledge management has been beneficial in organizing a firm's expertise so that it can be reused on future projects. Knowledge exists in both explicit and tacit forms. New web-based systems that make it possible for construction firm employees to interact and collaborate electronically can assist in the capture of tacit knowledge.

In Chapters 3 and 4, we will consider traditional methods used in the construction industry that have their focus on the organization of information. Chapter 6 will focus on project web portals that have their primary focus as a document repository for information but can also serve some knowledge management functions. Chapters 5 and 7 in this book will focus on technologies that can be used for knowledge management applications to capture a construction company's knowledge assets.

■ References

ASLIB, The Association for Information Management. 2003. http://www.aslib.co.uk/info/glossary.html (Accessed March 15, 2004; site now discontinued).

Alshawi, M. and Ingirige, B. 2003. Web-enabled project management: an emerging paradigm in construction. *Automation in Construction* 12: 349–364.

Canas, Alberto J., Carff, R., Hill, G., Carvalho, M., Arguedas, M., Eskridge, T.C., Lott, J., and Cravajal, R. 2005. *Concept Maps: Integrating Knowledge and Information Visualization*. Vol. 3426 of *Lecture Notes in Computer Science*, ed. Sigmar-Olaf, T. and T. Keller, 205–219.

Brooking, A. 1999. *Corporate Memory: Strategies for Knowledge Management*. London: Thomson Business Press.

Davenport, T. and Prusak, L. (1998) *Working Knowledge: How Organizations Manage What They Know*. Boston: Harvard Business School Press.

Davenport, T., DeLong, D., and Beers, M. 1997. Building Successful Knowledge Management Projects. Center for Business Innovation Working Paper, Ernst and Young, http://www.providersedge.com/docs/km_articles/Building_Successful_KM_Projects.pdf (Accessed October 1, 2005).

Egbu, C.O. and Botterill, K. 2002. Information Technologies and Knowledge Management: Their Usage and Effectiveness. *Electronic Journal of Information Technology in Construction* 7 http://www.itcon.org/ (Accessed March 15, 2004).

Jenkins, T. 2004. *Enterprise Content Management*. Waterloo, Ontario: Open Text Corporation.

New York State Department of Civil Service/Governor's Office of Employee Relations, 2002. Work Force Succession Planning-Tools & Resources: Knowledge Management/Transfer. http://www.cs.state.ny.us/successionplanning/workgroups/knowledgemanagement/terminology.html (Accessed September 30, 2005).

Nonaka, I. 1998 The Knowledge-Creating Company. In *Harvard Business Review on Knowledge Management*, Boston: Harvard Business School Publishing.

Norman, F. 1998. Definition of Information. http://www.eco.utexas.edu/Homepages/Faculty/Norman/long.extra/information/definitions.html (Accessed October 6, 2005).

Snyder, W.M. and deSouza Briggs, X. 2003. Communities of Practice: New Tools for Government Managers. IBM Center for the Business of Government. http://www.businessof government.org/pdfs/Snyder_report.pdf (Accessed February 20, 2006).

Tiwana, A. 2000. *The Knowledge Management Toolkit: Practical Techniques for Building a Knowledge Management System*. Upper Saddle River, NJ: Prentice Hall PTR.

CHAPTER 3

Using Computers for Construction Estimating

INTRODUCTION

It has long been recognized in the construction industry that estimating using the computer has many

benefits. Using computers reduces errors and reduces the time to produce estimates. Continuing

developments in information technology are serving to increase the capability of computer software to

automate the estimating process.

■ Estimating Software

Many different software packages are available to assist in developing estimates for construction projects. However, there are several exciting emerging technological improvements. First, there is an increasing trend to develop integrated software programs that combine estimating, accounting, and project management. For example, Timberline, which was originally for estimating, is now called Timberline Office and integrates accounting, document management, and job cost control with the core estimating functions.

Another trend is for various estimating software packages to easily exchange data with other software that performs different functions. For example, several estimating programs are able to exchange data with popular scheduling packages. Not only does an integration of this type allow estimating items to be associated with scheduling activities, it reduces the amount of time devoted to data input, because project data needs to be input only once.

Even more exciting are the emerging capabilities to automate the estimating process. It is now possible to generate estimate data automatically through the use of interoperability standards with CAD documents. CAD documents can be coded with detailed information about each project item that can be transferred to estimating programs. Additionally, programs are available that allow for automated quantity takeoff from electronic plan sheets. This provides the capability for a "paperless" estimating process with reductions in printing costs and time spent entering data manually.

Many estimating software packages are available. The cost of the estimating software varies from under $100 to many thousands of dollars. There are many software packages that have been developed for small construction projects and homebuilders. Well known among the software developed for small projects are Goldenseal, Bid4Build, and WinEstLT. Typically, the software for small projects does not have the sophisticated data exchange capabilities, nor can it integrate with other software packages that are contained in software conceived to handle large projects. Additionally, these software packages may be limited in their ability to develop the complex item coding schemes that may be used on a large project. However, these software packages are ideal for a small contractor because they are typically easy to learn and use.

For sophisticated projects, the different software packages tend to concentrate in particular segments of the construction market. For example, Timberline Office is a powerful estimating software that is most often used for estimating the construction of commercial building projects, whereas a software like HeavyBid is more suited to infrastructure projects and focuses on producing a line item, unit cost bid. Other well-known estimating products for large projects are Hard Dollar, MC^2, and Bid2Win. Bid2Win and Hard Dollar are oriented toward highway and infrastructure projects, whereas MC^2 is aimed at commercial construction. There is also estimating software aimed at government agencies and municipalities, as well as engineering firms for preparing engineer's estimates for transportation projects. An example of this type of software is Appia Estimator, which allows engineers' estimates to be prepared by government agencies and their design consultants.

■ A Discussion of Programs and Their Capabilities

This section will discuss some of the features of various estimating programs. All of the programs have powerful features that allow for data integration. The programs previously discussed all typically provide the capability to exchange data with Microsoft Excel. The use of spreadsheets is common in the development of estimates, and the ability to use standard estimating spreadsheets that a company has already developed is a useful feature. Most of the programs have the capability to export data to Primavera Project Planner and/or Microsoft Project. Both Timberline Estimating and Primavera are compatible with the IFC interoperability specifications (International Alliance for Interoperability). This allows for a very sophisticated linking between the two programs. Timberline Estimating includes a module called scheduling integrator that allows data from the Timberline estimate to be transferred to the Primavera scheduling program to automatically create scheduling activities. Activity durations are automatically generated from the estimate data. The generated activities can be automatically grouped in different ways, including by work breakdown structure, phase, or location (Timberline Office 2004c).

Some estimating software, including Timberline Estimating and MC^2, provide built-in modules that allow paperless methods of quantity takeoff by automatically calculating quantities from scaled CAD documents. Timberline Estimator has a module called On-Screen Takeoff. The software allows scaled drawings in several different CAD and graphic formats to be used to receive takeoff values and transfer them into Timberline estimating spreadsheets. The On-Screen Takeoff module allows lengths, areas, and volumes to be transferred for use in Timberline estimating. File formats that can be used include DWG, DXF, JPEG, BMP, TIFF, and PDF. The On-Screen Takeoff software also accepts other types of files including Dodge

Plan format, and government formats such as the Corps of Engineers. On-Screen Takeoff can also produce Excel files as output (Timberline Office 2004a).

On-Screen Takeoff can also be purchased separately for use with other estimating software packages. It is sold by On-Center software (On-Center). Figure 3-1 shows a landscaping CAD file that has been loaded into the On-Screen Takeoff Program image window. Various objects on the plan have been designated as different project items. In this case various plants, trees, and irrigation equipment have been designated. Figure 3-2 shows the On-Screen Takeoff window for the plan sheet. Here, items that have been selected in the image window are automatically tabulated. On-Screen Takeoff is capable of automatically counting, determining linear dimensions, and calculating volumes and areas.

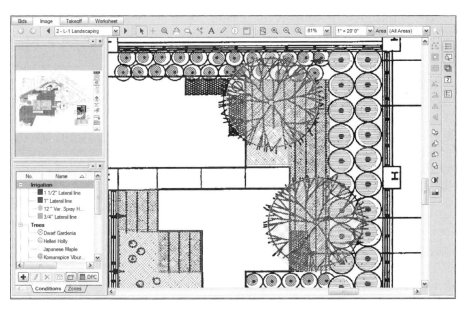

FIGURE 3-1   A Plan Sheet in the On-Screen Takeoff Image Window

The Capture Desktop feature of MC^2 provides similar functions and allows drawing to be sent in .jpg format attached to e-mails when there are questions about plan details. This can speed the exchange of information between the field and office or the designer when there are construction problems in the field (Management Computer Controls, Inc.).

Some of the estimating software includes standard cost database information. The Timberline estimating software includes databases of cost information, such as the RS Means building construction data. The inclusion of these databases allows a user to produce estimates rapidly even if they do not have their own historical data to draw on.

Software that focuses on commercial building construction tends to have some parametric estimating capabilities. MC^2 includes "estimating wizards" that allow conceptual estimates to be created. The wizards use a typical office building as an example and build a detailed estimate based on the input of the building parameters like the dimensions and the number of rooms. Timberline Estimating contains the Timberline Office Commercial

Bids | Image | Takeoff | Worksheet

No.	Name	Height	Area	Quantity 1	UOM1	Quantity 2	UOM2	Quantity 3	UOM3	Notes
⊟ Bed Plants										
5	Variegated Liriope	0"	(unassigned)	1,293	SF	321	LF	4	EA	Variegated Lir...
6	Big Blue Monkey Grass	0"	(unassigned)	1,403	SF	383	LF	5	EA	Big Blue Mon...
⊟ Ground Cover										
2	Mulch	0"	(unassigned)	3,761	SF	504	LF	2	EA	
3	Grass	0"	(unassigned)	276	SY	2,480	SF	196	LF	
⊟ Irrigation										
13	3/4" Lateral line	0"	(unassigned)	57	LF	8	EA	0	SF	3/4" Lateral li...
14	1" Lateral line	0"	(unassigned)	48	LF	8	EA	0	SF	1" Lateral line
15	1 1/2" Lateral line	0"	(unassigned)	21	LF	3	EA	0	SF	1 1/2" Lateral...
16	12 " Var. Spray Head	0"	(unassigned)	21	EA	21	EA	21	EA	12" variable s...
⊟ Miscellaneous										
1	Edge Trim	0"	(unassigned)	222	LF	106	EA	0	SF	
⊟ Trees										
7	Nellie R Stevens Holly	0"	(unassigned)	9	EA	9	EA	9	EA	
8	Japanese Maple	0"	(unassigned)	5	EA	5	EA	5	EA	
9	Peacock Orchid	0"	(unassigned)	21	EA	21	EA	21	EA	Peacock Orc...
10	Helleri Holly	0"	(unassigned)	36	EA	36	EA	36	EA	Helleri Holly 3...
11	Dwarf Gardenia	0"	(unassigned)	44	EA	44	EA	44	EA	Dwarf Garden...
12	Koreanspice Viburnum	0"	(unassigned)	1	EA	1	EA	1	EA	Koreanspice ...

FIGURE 3-2 Takeoff Window Showing Automatic Calculation of Quantities

Knowledgebase that contains pre-built models and assemblies of various construction systems (Timberline Office 2004b). The Commercial Knowledgebase allows a user to answer questions on a form and then a preliminary or detailed estimate is produced automatically. The purpose of the Commercial Knowledgebase is to allow users to prepare conceptual estimates quickly be eliminating the process of individual item takeoff. The user is prompted only for the data needed to calculate quantities and costs for assemblies or entire buildings.

Interoperability with CAD Documents

One of the interesting developments in the construction software industry has been the emergence of interoperability between various construction software packages. The International Alliance for Interoperability is an organization including members from the building, construction, and software industries. This organization has developed specifications for data exchange between computer programs, called Industry Foundation Classes (IFC). Right now the major construction software packages that provide interoperability are Timberline and Primavera Project Planner. Several CAD software packages, including AutoCAD by AutoDesk and MicroStation by Bentley, also conform to the interoperability standard.

Interoperability allows quantity takeoff directly from CAD documents into an estimating program if the CAD input data has been input using the IFC standard. Timberline Office CAD Integrator provides this capability as part of the Timberline Office software. This software allows the Timberline estimating software and IFC-compatible CAD files to be used in several sophisticated ways. For example, it is possible to highlight an object in the Timberline browser and see the same object highlighted in the drawing. It is also possible to click on the object in the drawing and see the associated costs in the browser.

Estimating software programs that focus on infrastructure projects are now featuring the capability to prepare estimates in a format compatible with online bidding systems, particularly those of the state departments of transportation. (Online bidding will be discussed in Chapter 8.) Both HeavyBid and Bid2Win can export completed line-item estimates to the online bidding system used by state departments of transportation.

■ Some Examples Using HeavyBid

HeavyBid is an estimating computer program that focuses on the heavy construction segment of the construction market. It focuses on the infrastructure market and provides five levels of the HeavyBid program depending on the size of the project. HeavyBid has been used on projects from $50,000 to over $1 billion. To illustrate the functions of estimating software and to show how a bid is prepared, several examples from the HeavyBid program will be presented. Examples that outline how a bid is prepared using HeavyBid are presented.

The Master Estimate

A master estimate can be established for use with HeavyBid. In this master estimate the various types of labor, equipment, materials, and subcontractors can be defined. The information entered in the master estimate is information that is used repeatedly on the construction firm's project. Thus the master estimate can be used as the basis for all new project estimates. HeavyBid also allows any other existing estimate to be used as the basis for a new estimate. If the construction company has just purchased a new asphalt paving machine for use on its projects we can add it to the master schedule by clicking on the equipment icon. This is the icon with a picture of a bulldozer. Clicking on this icon opens a window where we can input the new asphalt paving machine and its charging rate. Figure 3-3 shows how the new machine is added to the master estimate in HeavyBid.

Establishing a Bid

HeavyBid takes a hierarchical approach to establishing the project cost. HeavyBid develops a bid price for each bid item that is defined for a project. Infrastructure projects are typically unit price and require a unit price and total price to be developed for each bid item. Each bid item is composed of activities that consume resources. Each activity requires various resources such as labor, equipment, materials, and subcontractors to complete. These various resources and their costs are associated with each activity, and the activity costs are aggregated to develop the bid item price.

HeavyBid has the capability of reading in standard bid items from DOTs, or the bid items can be entered manually. Figure 3-4 shows the list of bid items for an asphalt paving project that are listed in a tree structure at the left of the screen. If a bid item is clicked, the tree expands to show the list of activities incorporated in the item. If an activity is selected, the tree expands further to show all of the resources required for the activity. In this case we see a listing of the equipment and labor required for asphalt paving. Figure 3-5 shows the asphalt paving bid item completely expanded, showing all the activities, and then a listing of the resources required for an activity.

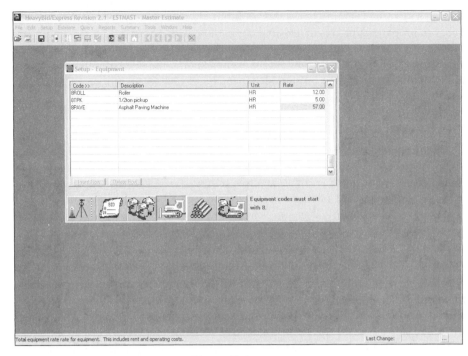

FIGURE 3-3   Adding Equipment to Master Estimate

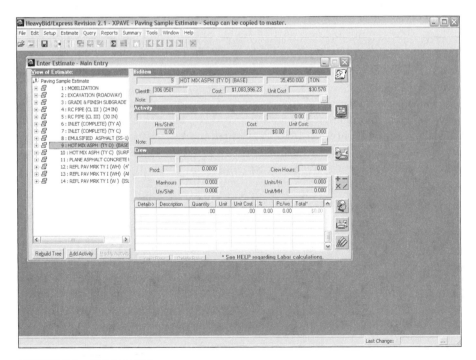

FIGURE 3-4   Bid Item Tree

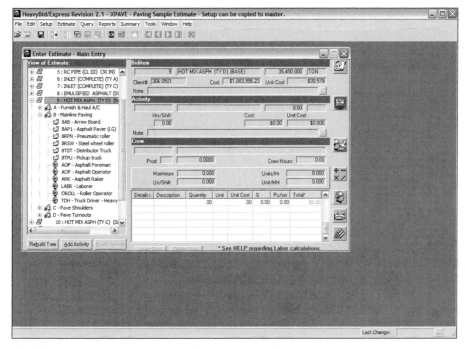

FIGURE 3-5 Expanded Tree Showing Activities and Resources

Adding Bid Items, Activities, and Resources

It is often stated that construction is the conversion of 4Ms—Materials, Manpower, Machinery, and Money—into a completed project (Halpin & Woodhead 1998). In HeavyBid the project is broken up into a hierarchy of items and activities that consist of the required materials labor and equipment for the project. This hierarchical approach assists in the identification of the required resources. A bid item is decomposed into its constituent activities. This allows the estimator to then identify the required resources for each activity.

Additional bid items can be easily added to an estimate. Bid items in HeavyBid are numbered. The user is required to input the quantity and unit for the bid item. After adding a bid item the user goes to the main entry screen for the bid and selects the new bid item. To add an activity to the bid item, users click on add activity in the lower left of the screen and the add activity window appears. This is illustrated in Figure 3-6. The HeavyBid instructions indicate that activities should be consecutively lettered for small projects. Therefore, the activity is given the designation A. Any name can be input for the activity, and it has been decided to call it Pave Curbs. We could use the predefined crews to define this activity, but instead we will select the required resources from the lists of defined resources.

Equipment, labor, and material resources can be easily added to an estimate in HeavyBid. These windows are activated by clicking the icons at the right of the estimate window. The construction worker icon is for labor, and the bulldozer icon is for equipment. Selecting the icon resembling pipes brings up the predefined materials list. In this case we select hot mix asphalt, and it is brought into the activity. Figure 3-7 shows the final step of entering the material information, where the unit price for the asphalt is entered in the activity window.

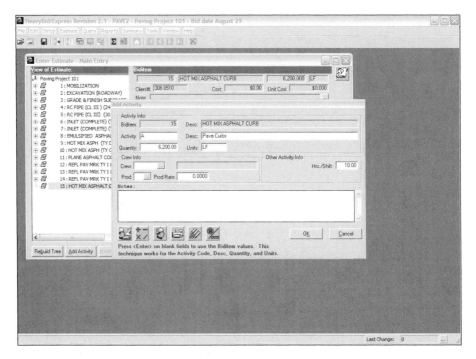

FIGURE 3-6   Adding an Activity

Finally, to complete the addition of the activity, we adjust the man-hours required for the labor and equipment based on our knowledge of the time required to add the bid item to the estimate. This is shown in Figure 3-8. Note that all of the details of the added resources are shown in the spreadsheet at the bottom right of the screen. It should also be noted that the Crew section of the activity is blank. If we had used standard crews, this information would be filled in. It is important to note that the productivity of a crew is defined in this window. Various units for productivity can be selected from the productivity code window. These include units/hour, units/shift, units/man-hour, man-hours/unit, $/unit, and crew-hours.

The HeavyBid program also allows new resource information to be inserted directly into the activity spreadsheet without the need to predefine the resources. For example, an equipment rental not part of a company's equipment fleet could be entered directly for a specific project.

Finishing the Bid Estimate

Entering all of the bid items and their associated activities produces the contractors cost to perform the work. Of course, the bid estimate submitted to the owner must include markups for overhead and profit. A strength of HeavyBid is its ability to produce the marked-up costs that are necessary for the unit price type of contracts seen in infrastructure projects.

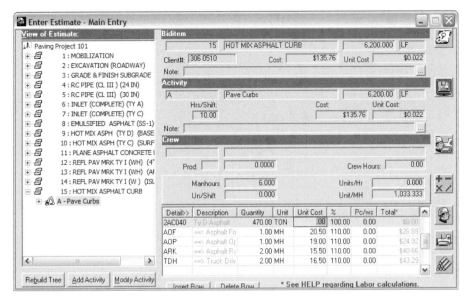

FIGURE 3-7 Entering Material Unit Price

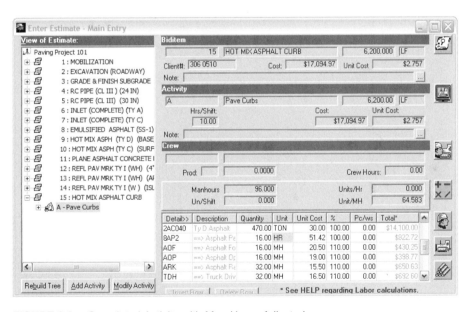

FIGURE 3-8 Completed Activity with Man-Hours Adjusted

Figure 3-9 shows the bid summary screen. Here a contractor may determine a separate percentage markup for each type of cost. Additionally, a contractor may have certain costs that he wishes to spread over all project items. By clicking the Enter Addons tab of the bid summary it is possible to add these additional costs. Figure 3-10 shows the addons for

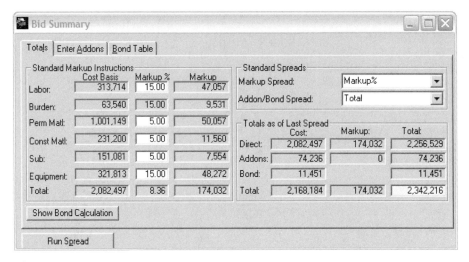

FIGURE 3-9   Bid Summary Screen

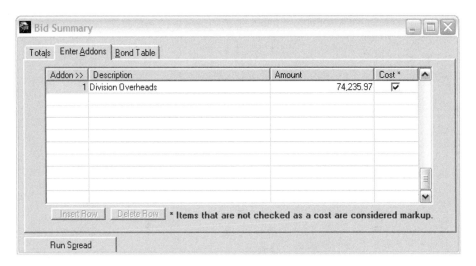

FIGURE 3-10   Project Addons

this project. In this case overheads are going to be added to the project cost and spread over all of the bid items total direct costs. Additional options are also possible. The other options include spreading the cost over the labor costs only, to prorate the costs over total costs less subcontractor costs, or no spread. If no spread is selected the additional costs must be distributed by using manual bid unbalancing. The markup is spread back to the bid items in a similar manner. To calculate the total contract cost after markups and additional costs are included, the user must click the "Run Spread" button. The total project cost is calculated in the bottom right-hand corner of the bid summary screen.

The cost of the bid bonds is also incorporated in the bid at this point. By clicking the Bond Table tab it is possible to view and input the bonding costs associated with the project. Typically, bond prices are quoted as a percentage of the contract amount, and this table allows these percentages to be easily input.

Figure 3-11 shows the bid pricing window where the bid price for each bid item is displayed. This is the final stage of the bid preparation process using HeavyBid. This screen shows a balanced price, bid price, and bid total for each item. Each bid item has a balanced price; this is the unit price for the bid item including all direct costs and additional overhead costs. The bid price is the same as the balanced price, but can be modified manually by the estimator. The bid total column is the bid price multiplied by the item quantity. Therefore, it can be observed that the HeavyBid program automatically takes the raw cost and productivity input, and transforms it automatically into the required unit prices and totals that would be required to fill out a proposal form for an infrastructure project bid.

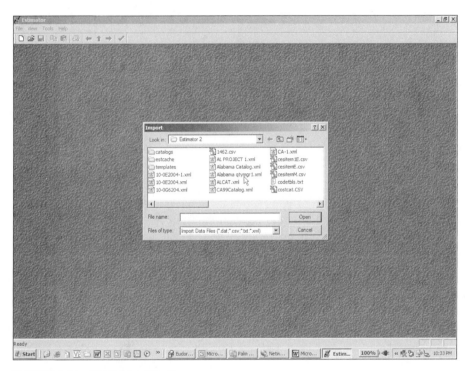

FIGURE 3-11 Bid Pricing Window
Courtesy of: Joe Phelan, Director, Info. Tech, Inc.

HeavyBid also has the capability of checking for bid errors. Figure 3-12 shows the estimate inquiry screen that is accessed from the query menu. In this screen possible bidding errors are highlighted. For example, this screen will display the number of items where the bid price is less than the calculated cost. The screen also flags bid items that are zero, probably indicating a bid item that has not been populated with cost data and is blank.

Using Computers for Construction Estimating

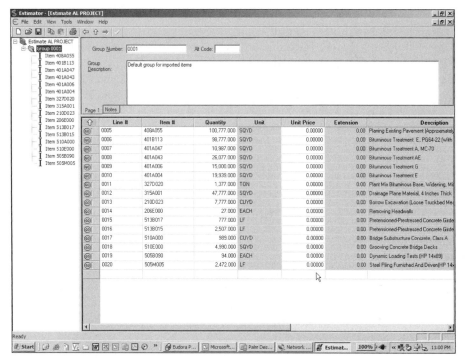

FIGURE 3-12 Estimate Inquiry Window
Courtesy of: Joe Phelan, Director, Info. Tech, Inc.

Exporting and Importing HeavyBid Data

HeavyBid can import proposal and bid item information from the online bidding service used by most state departments of transportation. When a bid is complete, HeavyBid has the capability to export the bid data in a format that allows it to be submitted to the online bidding system without reentry of data. HeavyBid can also import and export data with the Excel spreadsheet program.

■ Appia Estimator: Estimating Software for the Owner and Designer

Appia Estimator by Info Tech is an estimating program that has a different emphasis than programs designed for construction contractors preparing a bid. Appia Estimator is a tool aimed at owners and designers of highway and infrastructure projects. The Appia Estimator system is used by 23 state Departments of Transportation. Appia is designed to work with the standard specifications and master item list developed by most DOTs for their unit price infrastructure projects.

Typically, the Appia Estimator software is used by design consultants and DOTs to generate the engineer's estimate. The use of a master item list allows Appia to develop an engineer's estimate based on historical bid tabulations of the agency's past projects. Naturally, an owner's perspective is different from a contractor who is attempting to achieve a competitive price while maintaining an adequate profit. In generating an engineer's estimate using Appia Estimator a state DOT is seeking to produce a reasonable figure that can be compared with the contractors' bids.

Appia Estimator has several features that reduce the time it takes an agency to produce an engineer's estimate:

- The program is able to accept data files containing quantities for project items from CAD drawings produced using Bentley Microstation. (CAD is discussed in more detail in Chapter 9.) Quantities are calculated from Microstation CAD drawings using a program called Bentley Quantity Manager. Bentley Quantity Manager produces a file of quantities in XML format that can be read by the Appia Estimator program. XML (Extensible Markup Language) files are a type of file that is used for data exchange between programs. Figure 3-13 shows the Appia Estimator file import window. Figure 3-14 shows the Appia Estimator program after the data from the XML file has been loaded.

- Most DOTs maintain a catalog of average historical bid prices for standard cost items. The Appia Estimator program allows these standard cost catalogs to be automatically applied to the item and quantity information. Figure 3-15 shows how the unit prices are automatically generated after a cost catalog is applied to an estimate.

- An interesting feature of Appia Estimator is the capability to modify the unit prices based on historical records of the relationship between the quantity and unit price for a particular project bid item. Due to economies of scale, unit prices for larger quantities are typically lower than for projects in which only a small quantity of work is performed. Appia Estimator has a graphical display that shows the relationship and allows a user to modify the unit price based on the quantity of an item to be installed. This is shown in Figure 3-16.

- Figure 3-17 shows that the Appia Estimator program has the capability to open a web browser window and access standard labor rates from a U.S. Department of Labor web site for use in estimating labor costs.

- The Appia Estimator program can also be linked to the Bid Express web site used by many DOTs to access item and cost data. This is shown in Figure 3-18. The use of the BidExpress web site is discussed in detail in Chapter 8.

Appia Estimator differs from the other estimating programs discussed in this chapter because its primary purpose it to create an engineer's estimate for use by a government agency. It illustrates how various estimating tools are tailored to an organization's particular needs.

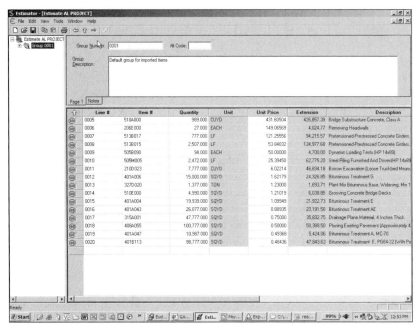

FIGURE 3-13 Appia Estimator File Import Window
Courtesy of: Joe Phelan, Director, Info. Tech, Inc.

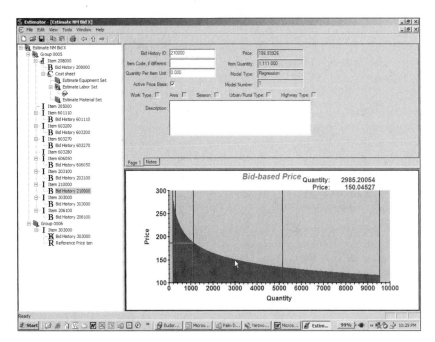

FIGURE 3-14 Estimator with Imported Quantities
Courtesy of: Joe Phelan, Director, Info. Tech, Inc.

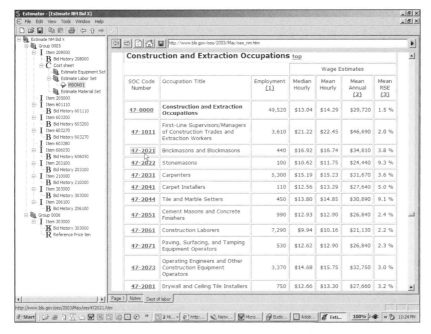

FIGURE 3-15 Estimate with Unit Prices Input from a Cost Catalog
Courtesy of: Joe Phelan, Director, Info. Tech, Inc.

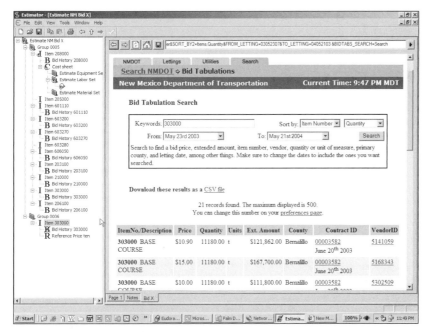

FIGURE 3-16 Screen Showing Capability to Modify Unit Prices Based on Quantity
Courtesy of: Joe Phelan, Director, Info. Tech, Inc.

Using Computers for Construction Estimating

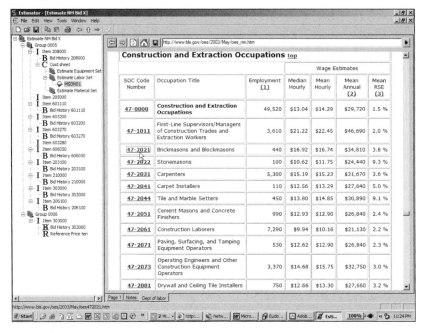

FIGURE 3-17   Accessing Labor Costs

Courtesy of: Appia® Estimator™ by Info Tech, Inc.

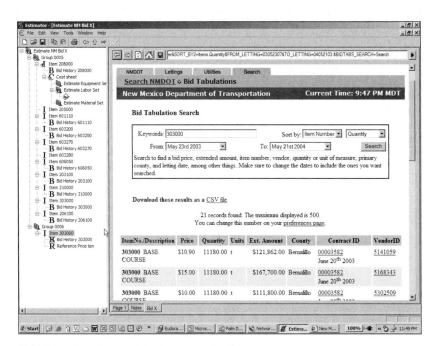

FIGURE 3-18   Getting Cost Data from BidExpress

Courtesy of: Appia® Estimator™ by Info Tech, Inc.

Chapter 3

■ Some Conclusions About the Use of Estimating Software

This review of available software, and the HeavyBid and Appia Estimator examples, illustrate the benefits of using estimating software. First, the estimating software automates many of the time-consuming tasks involved in preparing an estimate. Using the error-checking features of the software can eliminate mistakes. Second, the estimating software performs some calculations that could be difficult and time consuming. For example, the HeavyBid software is able to calculate a unit price for each bid item in the required units when the inputs provided by the user for the activities may be in different units.

This chapter also highlighted the fact that continuing developments in estimating software are moving toward integration with other software, such as CAD documents and scheduling software. There will be a continuing increase of the capability to exchange data between software, and as increasingly more software developers adopt interoperability standards, the ease of exchanging data between programs will increase dramatically. These developments are good for construction contractors because they will allow project data to be input once and then reused in various ways, reducing management costs and increasing workers productivity. It has also been shown that automated, paperless quantity takeoff is now a reality.

■ Web Links To Estimating Software

Table 3-1 provides links to the estimating and quantity take-off computer software discussed in this chapter.

TABLE 3-1 Estimating software web links

Estimating Software	Web Page	Comments
Build4Bid	www.build4bid.com	Small projects
Goldenseal	www.turtlesoft.com	Small projects
WinEst	www.winest.com	Commercial construction
MC^2	www.mc2-ice.com	Commercial construction
Timberline	www.timberline.com	Commercial construction
Hard Dollar	www.harddollar.com	Heavy construction
Bid2Win	www.bid2win.com	Heavy construction
HeavyBid	www.hcss.com	HeavyBid
Appia Estimator	www.infotechfl.com/software_solutions/estimator.shtml	Estimating for owner agencies
On-Center On-Screen Takeoff	www.oncenter.com/products/ost.asp	Quantity takeoff software

■ References

International Alliance for Interoperability. Technical-Industry Foundation Classes. http://www.interoperability.org.au/608.html (Accessed September 22, 2005).

Management Computer Controls, Inc., Estimating Wizards. http://www.mc2-ice.com/ice_prod_frame.htm (Accessed September 25, 2005).

On-Center Software, On-Screen Takeoff. http://www.oncenter.com/products/ost.asp (Accessed September 26, 2005).

Timberline Software. 2004a. ePlan Takeoff. http://sagetimberlineoffice.com/include/pdfs/ePlan_takeoff.pdf (Accessed September 26, 2005).

Timberline Software Corp, 2004b. Model Estimating. http://sagetimberlineoffice.com/include/pdfs/model_estimating.pdf (Accessed September 26, 2005).

Timberline Software Corporation, 2004c. Scheduling Integrator. http://sagetimberlineoffice.com/include/pdfs/scheduling_integrator.pdf (Accessed September 26, 2005).

CHAPTER **4**

Scheduling and the Computer

INTRODUCTION

Scheduling software is a primary information technology application in construction. With the advent of the personal computer, scheduling software is widely available to contractors of all sizes. Computer software used for construction scheduling and planning typically employs the Critical Path Method (CPM). For simple projects, contractors may occasionally use simple bar charts created in spreadsheet programs (see Table 4-1 for descriptions of the major scheduling software). For CPM scheduling, there are several software packages that are widely used in the construction industry. These include:

- Primavera

- Microsoft Project (Standard and Professional)

- Primavera Suretrak

- Primavera Contractor

Although no comprehensive study has been conducted concerning how these software packages have been employed in the industry, it has been reported (Newitt 2005) that Primavera is preferred in the construction industry for complex projects, because of its ability to integrate cost control with scheduling activities.

TABLE 4-1 Types of Software used in the construction industry

Software	Market software is typically used in	Cost of Software (single user version)
MS Project Standard	Smaller firms, although some large organizations prefer it for managing projects	$600
MS Project Professional	Small to Large Projects	$1,000
Primavera	Large, Complex Projects	$4,000
Primavera Suretrak	Small- to medium-sized projects	$600
Primavera Contractor	Small- to medium-sized projects	$600

Microsoft Project also has the benefit of integrating with the Microsoft Office suite of programs that is particularly useful to small contractors. For both Primavera and Microsoft Project, there are additional software costs for the server versions that allow the software to be used by multiple users on a LAN, and to establish web-based access to schedule data. There are several high-end scheduling software packages that are very expensive, and are most commonly used in the construction industry for the simultaneous scheduling of multiple complex projects. A good example of this type of software is Welcom OpenPlan. This program is also used in the aerospace industry to manage complex projects. This program resides on a server. Its main advantage is its ability to handle large numbers of interrelated projects and users.

Industry Acceptance of CPM Software for Scheduling

Using the computer to schedule construction projects is perceived as a very important activity within the construction industry. A study of CPM software use by the ENR Top 400 Contractors has found that these large contractors generally consider the use of CPM as important to their companies' success (Kelleher 2004). The study indicated that:

- Of the companies studied, 98% feel it is a valid management tool.
- 80% believe it increases communication within the workforce.
- Top 400 Contractors use CPM as a project-planning tool before construction and to periodically update schedules during construction.
- Contractors are increasingly using CPM during the estimating and bidding phases of a project to improve their understanding of the sequence of proposed project activities.

■ Simple Schedules Using Excel

Figure 4-1 shows an Excel spreadsheet created by a construction manager for the construction of a large school building. This Excel spreadsheet demonstrates how to create a bar chart construction schedule for project planning and display to the project owner.

In this schedule the construction manager has:

- Divided the schedule into several blocks representing the major segments of the project such as construction of an auditorium and construction of a cafeteria.
- Used different color bars for each activity in a block by selecting the cell color for each range of cells in a bar.
- Schedule can be easily updated by changing cell colors as appropriate.

This Excel bar chart illustrates a simple way in which computers can be used to plan and schedule construction projects. This type of schedule is appropriate for small projects for which there are relatively few activities and the sequence of construction activities is well defined.

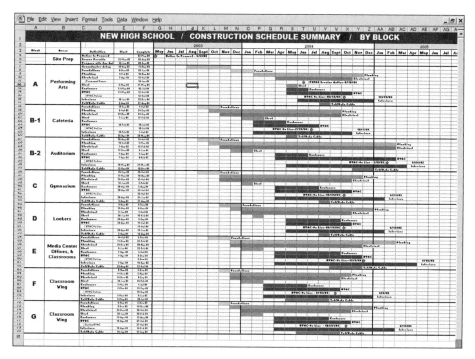

FIGURE 4-1 Schedule Bar Chart Using Microsoft Excel

■ Critical Path Method Scheduling and the Computer

By far the most popular technique for scheduling construction projects is the Critical Path Method. Excellent software is available for personal computers enabling its widespread application in the construction industry. This section describes the software that is typically used and discusses the ways CPM can be employed to plan and schedule construction projects.

Introduction

The Critical Path Method (CPM) is widely used in the construction industry. The application of CPM on the computer allows schedules for complex projects to be efficiently generated. The use of CPM on the computer also allows for consideration of the interactions between activities, and the ability to schedule with constrained resources. The widespread availability of CPM software for personnel computers and the ability to acquire it at relatively low prices allows a broad range of contractors to employ it.

Applying CPM on the computer provides many benefits. Today's projects have become increasingly complex and it would be difficult to schedule a CPM project using pencil and paper calculations. Additionally, complex projects are subject to many changes, and delays. The computer allows CPM schedules to be rapidly changed and updated.

Computer software now in use incorporates many powerful and useful features, including automatic generation of bar charts, CPM diagrams, and reports. All available CPM software can provide output in terms of calendar dates, and account for weekends and holidays when calculating the schedule.

Reasons for Using CPM to Schedule Construction Projects

CPM and its applications to construction project management have been widely discussed in the literature. There are many books that describe how to use CPM and also how to use the many features of the available software. There are several reasons for the popularity of CPM construction. Some of the most important reasons for using the CPM technique are:

- Improved advanced planning and scheduling of construction projects, as well as tighter control of project status during construction provides CPM users with the potential to reduce total construction time.
- Computerized CPM allows projects to be better organized and can increase labor productivity.
- CPM allows a well-planned schedule to be developed that provides for more uniform use of resources throughout the project.
- Compiling data for a CPM schedule forces contractors to do detailed thinking and planning about how the project is to be conducted.
- CPM computer software improves project communications by providing all of the project participants with detailed, graphical reports of the project status, and also provides other participants, such as the owner, with information about a contractors planned activity sequence (Newitt 2005).

What is the Critical Path Method?

The Critical Path Method (CPM) has been used in the construction industry since the 1950s. Its use is closely associated with the development of computer hardware and software. CPM is a method that does not employ any advanced mathematics, but without computers its use would be too cumbersome for the complex projects that are now commonplace.

Defining the Computer Input for A CPM Analysis

A good understanding of CPM is necessary to properly employ CPM-based scheduling software. CPM represents the activities in a construction project as a set of interconnected nodes. Each node in the network represents an activity. Typically, CPM software uses the activity on node method of representing a network. It is up to the scheduler to decompose the project into appropriate activities and construct the network. In constructing the network the scheduler must understand the relationship between the various activities in the project, and understand the dependencies between activities.

Defining the durations of activities is a vital component of developing a CPM schedule. Typically, construction companies use a mixture of records from similar past projects and their experience to produce a duration estimate for an activity. Using the CPM method, durations are always input as constants.

The Critical Path Calculations

Many excellent books have been written about CPM scheduling (Hinze 2003; O'Brian 2005; Weber 2004). The author refers readers to these books for details of the CPM calculations. In this book we will offer a general outline of what is produced by a computer software program performing the CPM calculations on a network.

When using CPM, a total project duration is calculated. The sum of the durations of the longest path through the network constitutes the project duration. This longest path through the network is called the critical path. Activities not on the critical path may have some scheduling float (sometimes called slack or leeway).

Activities on the critical path must start as soon as the previous critical activity is completed or the project duration will increase. Therefore activities on the critical path are called critical activities and have no scheduling leeway. Activities not on the critical path may be delayed and not affect the total project duration. A contractor, therefore, has some discretion as to when non-critical activities may start. CPM programs typically calculate an early start and late start for non-critical activities.

Scheduling Float

Although there are several types of scheduling float that can be calculated, total float is typically the only type of float that is considered on construction projects. Total float is defined as the amount of time that an activity can be delayed and not delay the completion date of the project, although it may delay the start of subsequent activities. All of the popular scheduling programs calculate the total float for non-critical activities.

■ The Capabilities of CPM Software

This section addresses the capabilities of CPM software. First we consider the basic capabilities to perform CPM calculations and produce schedule reports. Then we will discuss the powerful capabilities of the software to link the schedule with resource usage and cost.

The Core Functions of Any CPM Software

The core function of any CPM software is to calculate the project duration, identify the critical activities, and identify the total float for non-critical activities. Importantly, with the continuing development of computer graphics and printing capabilities, CPM programs now offer many different types of outputs that provide a clear representation of the schedule that are easily used by various types of users. Most scheduling programs now provide output of the scheduling information as color-coded bar charts, time-scaled network diagrams, and tabular reports. The software allows information to be tailored to various types of users, such as reports for top management, and simple bar charts that are understandable to users at the construction job site.

A summary of the basic features of CPM software includes:

- Input forms to name activities, code activities, input durations, and the relationships between activities.
- Calculation of projection duration. The CPM software calculates the completion date of the project.
- Determination of critical activities. Critical activities are flagged. Typically program output clearly identifies critical activities for the software user.
- Calculation of floats for non-critical activities. Non-critical activities are identified. Programs have controls that allow users to select when non-critical activities start (such as Early Start or Late Start).
- Calendar Management, including the ability to account for work stoppages such as weekends and holidays and to maintain multiple calendars.
- Production of high-quality reports and charts to disseminate the schedule information including bar charts, and other types of reports. Programs typically have the capability to print bar charts in color, and to print high-quality versions of the project network.
- Allows for users to update activity details and recalculate the schedule details. CPM programs typically allow different "versions" of the schedule to be maintained so that the original project plan can be compared to the actual project schedule.

Some examples of how data are input to a scheduling program are shown in Figures 4-2, 4-3, and 4-4. Figure 4-2 shows how data can be input using the Gantt chart view of Microsoft Project. Users input an activity name and duration. Highlighted activities can be linked together using the link icon on the toolbar. The project Gantt chart is displayed on the right of the screen and updates as changes are made to the scheduling activities. Figure 4-3 shows how resource data can be added in Microsoft Project. A tabular form is used to input the data. The user can select the columns to be displayed and can create custom data columns. Figure 4-4 shows the Network Diagram view for Microsoft Project. In this view new activities can be created, and activities can be linked by dragging an arrow between two activities. This provides a rapid way of inputting the relationships between activities.

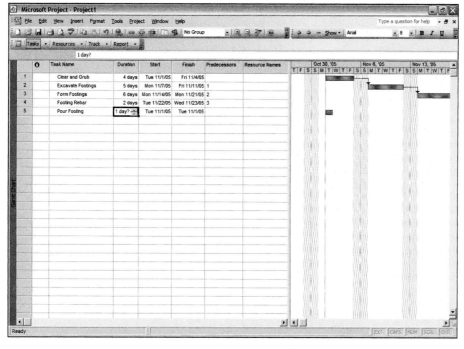

FIGURE 4-2 Inputting Information in a Microsoft Project Gantt Chart

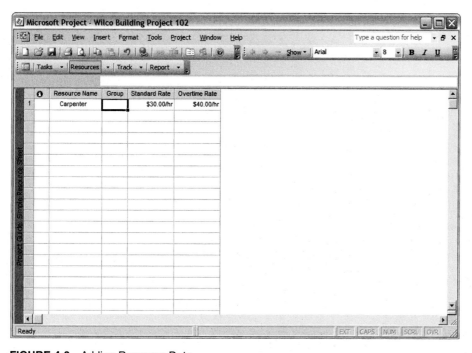

FIGURE 4-3 Adding Resource Data

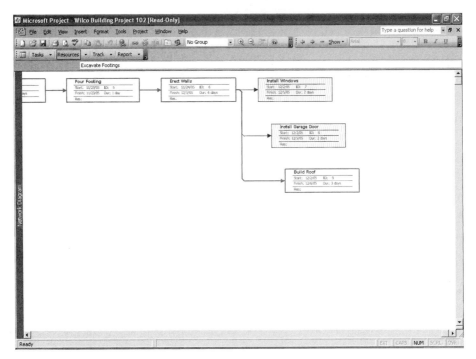

FIGURE 4-4 Network Diagram View in Microsoft Project

An example of the excellent bar charts produced by CPM scheduling programs is displayed in Figure 4-5 that has been produced in Primavera. This figure shows a portion of a bar chart produced for an actual process plant retrofit. In the figure, the wide bars indicate the actual schedule. The thin black bars show the original project plan. A careful examination of the bar chart shows that the sequence of activities has been modified during construction. Using Primavera, the same information in the bar chart can be viewed in network form.

Figure 4-6 shows a Primavera PERT view. This is actually a network drawing showing the interrelationships between activities, duration, and total float. Those activities with a slash through them have been completed. These figures illustrate the capability of scheduling software to communicate schedule information. The potential of scheduling software to provide high-quality reports for analysis is shown in Figure 4-7. The report shows cost, schedule, and resource data for multiple projects.

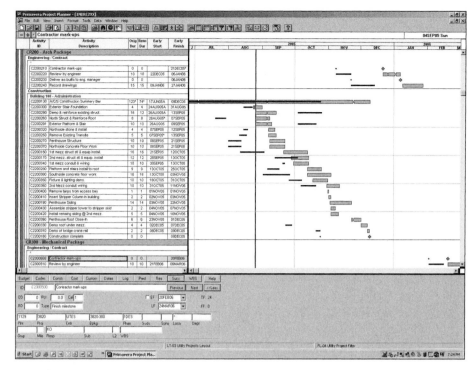

FIGURE 4-5 Bar Chart from a Process Plant Retrofit

Scheduling and the Computer

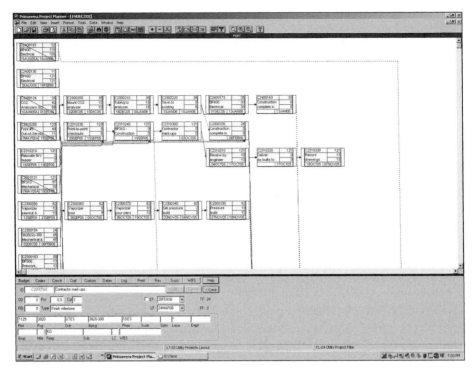

FIGURE 4-6 Primavera Pert View of Example Project

Advanced Scheduling Capabilities Using CPM Software

Scheduling software has many advanced features. In particular the ability to associate cost data with scheduling activities can provide improved cost control during the construction process. Additionally, scheduling programs provide the capability to consider resource limitations in the calculation of the schedule and to level resource usage during a project.

Some of the most important features that an advanced user can employ when using scheduling software are:

- Ability to have a "cost-loaded" schedule in which each activity has a cost assigned to it.
- Construction budgeted versus actual costs can be compared.
- CPM software is typically able to generate S-curves of project expenditures.
- Project resources such as labor and equipment can be associated with each scheduling activity.
- Scheduling software has the capability to level the use of multiple resources and to minimize the occurrence of undesirable spikes in resource usage.
- Major scheduling software programs employ precedence diagramming to build the network (activity on node networks), allowing for the definition of different scheduling relationships other than the basic finish-to-start relationship.
- Relationships like start-to-start, finish-to-finish, and start-to-finish are possible.

FIGURE 4-7 Primavera Report Combining Data from Several Projects
Courtesy of: Primavera Systems, Inc.

- Leads and lags are possible.
- Ability to use different scheduling logic can provide increased flexibility when defining the schedule, but can also make the schedule difficult to interpret.
- Various activity codes can be assigned to scheduling activities to organize the activities into groups.
- Typical activity code types in responsibility, area, phase, and user may define additional codes to activities.
- Ability to consolidate related projects into a master project.
- Uses of this feature include: ability to schedule different phases of a project as subproject, ability to schedule projects where work takes place at several different geographical sites, and each site can be scheduled as a separate subproject.
- Resources can be gathered in common pools that are shared among the various ongoing projects, allowing for resource limitations to be accurately represented in developing schedules.
- Programs such as Primavera and Microsoft Project can perform earned value analysis that is a method for measuring project performance, allowing for analysis of budget spent, amount of work completed, and the original budgeted amount for the work.
- Data exchange with other types of programs, particularly estimating software to allow budget information to be downloaded to scheduling activities from the estimate without the need to input data twice.

Both Primavera and Microsoft Project have many features. These features give these programs the capability to adapt and to be customized to almost any situation encountered on a project. However it can take considerable effort on the part of a scheduler to learn all of the nuances of these programs. It is suggested that any contractor seeking to implement these programs ensure that staff receives proper training.

Multi-User and Web-Based Scheduling

For large companies involved in complex construction projects, many different people may be involved in developing and updating schedules. There may be many other people who want to receive the latest schedule updates and reports on a project, such as top management wishing to assess the status of a project. Therefore, the scheduling software from Primavera and Microsoft can operate on a corporate intranet, allowing multiple users to access and modify the same project schedule and view output. Additionally, these computer programs provide the capability to access via the web. To provide these capabilities requires a user to set up a computer server and acquire server and database software. Most construction companies would need to hire professional IT assistance to implement an intranet, but the potential benefits are great and even relatively small construction companies should consider implementing an intranet for scheduling.

Requirements for Using Microsoft Project Software in a Client/Server Environment

To implement Microsoft Project on an intranet requires several programs in addition to the installation of Microsoft Project on each computer:

- Microsoft Server software, Microsoft Project Server, Microsoft SQL Server, and Microsoft Windows SharePoint Services are required.
- SharePoint Services is required for project collaboration because it is the software that allows multiple users to share Project schedule files, and allows individual users to check-in/check-out files as they modify them.
- SharePoint Services has many interesting features. Its Content Management System features are discussed in detail in Chapter 7.

Enabling Web-Based Exchange of Information Using Microsoft Project

Beyond Client/Server networks it is possible for users to access Microsoft Project via the Internet. This enables managers at remote locations to access schedule information. To enable Internet access requires some additional steps, including:

- Requires a user to provide a server, and to use Microsoft Project Professional 2003, Project Server 2003, and Microsoft Project Web Access 2003 (Microsoft Corporation)
- Requires a server running the Project Server software
- Each computer with web access must have the Project Web Access software installed

Internet access using Microsoft Project has several beneficial features. These include:

- Managers can assign tasks to team members and receive reports of work that is completed from users on the web.
- Reports of progress from the field can be received and consolidated status reports can be produced by the software.

Configuring Primavera for Networks and Web Access

Primavera can be configured in several ways to work over an intranet, and it can be configured to allow user access over the Internet. To use Primavera on an intranet it is necessary to set up a client/server database to store the project data. To use Primavera in an intranet environment requires database software to be installed on the server. Primavera runs with Oracle, Microsoft SQL server, or Microsoft SQL Server Desktop Engine.

To implement web-based access for Primavera requires:

- Additional server-based components must be installed.
- The primary component is called myPrimavera. This application provides web-based access to project, portfolio, and resource data across an organization.
- Every myPrimavera user can create a custom workspace to view information about specific projects and categories in which they may be interested (Primavera 2005). A typical myPrimavera screen for an executive viewing the status of several projects simultaneously is shown in Figure 4-8.

■ Monte Carlo Simulation—An Extension of CPM

PertMaster is an interesting program that accepts data from both Primavera and Microsoft Project that can provide further schedule analysis of the CPM schedule by considering the variations possible in planning the duration of scheduling activities and activity costs. Pert-Master can help construction managers assess the risk inherent in a schedule. The basic CPM technique considers durations to be constant values. The program allows activity durations and costs to be input as statistical distributions to reflect the potential variability during a construction project.

PertMaster uses Monte Carlo simulation that involves performing a large number of trial runs called simulations, and calculates the probability distribution of possible outcomes. In this case, the probability of particular schedule durations can be determined. PertMaster samples from the statistical distributions defining each activity. It then performs the CPM calculations and determines the projects duration and stores the output. It then repeats these two steps many times with many possible combinations of the activity durations. The output from the program is a histogram of frequency of the simulated project durations. The software also generates the probability of completing the project on schedule using deterministic durations, and generates statistics about the mean, median, and maximum durations. Statistics can also be viewed for individual activities. PertMaster can simultaneously simulate the activity costs and produces similar outputs concerning the probability of achieving various total project costs.

FIGURE 4-8 MyPrimavera: Web-based Access to Schedule Information
Courtesy of: Primavera Systems, Inc.

PertMaster has an interface similar to most scheduling programs that can be used for stand-alone scheduling using deterministic durations. However, its primary usefulness is its simulation abilities and its capabilities to accept input, both duration and cost, from other scheduling software. The simulation capabilities of PertMaster allow construction planners to consider the risk and likelihood of meeting project schedule and cost goals. The following section presents an example of using PertMaster to study the effects of schedule and cost variations.

■ A PertMaster Example

This example shows how Monte Carlo simulation can be applied to a typical building project schedule using PertMaster. Figure 4-9 shows a schedule that has been entered with the bar chart view displayed. Each scheduling activity has been defined as a triangular distribution. The durations are shown in the right-hand columns. The bar chart displays the schedule using the most likely duration.

Figure 4-10 shows the schedule with the task details window displayed. Selecting and clicking on an activity opens this window. In this window the user can select a probability distribution for an activity. Various statistical distributions can be used to define probabilistic activity durations including the beta, lognormal, normal, triangular, and uniform distributions. In Figure 4-10, a triangular distribution has been input for the activity.

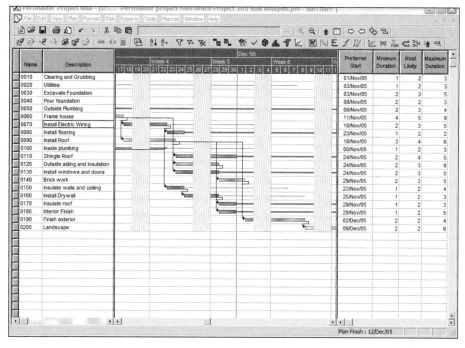

FIGURE 4-9 PertMaster Bar Chart

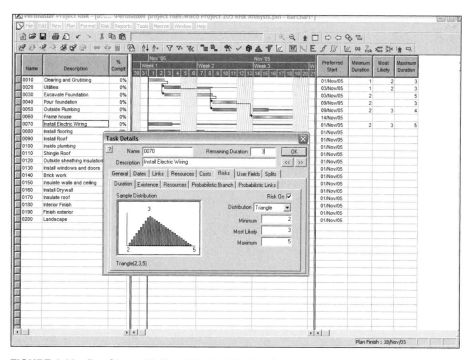

FIGURE 4-10 Bar Chart with Task Window Displayed

To perform a risk analysis requires the user to specify the number of iterations. The program default setting is 1000 iterations. That is, the software samples from the statistical distributions and calculates project duration and cost 1000 times. It takes a few seconds to run all of the simulations on a PC. Figure 4-11 shows the various options available when running a risk analysis.

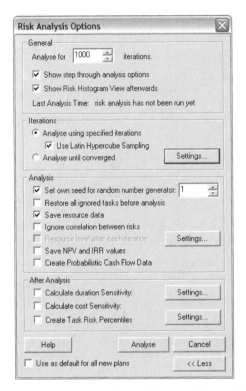

FIGURE 4-11   PertMaster Window to Select Simulation Options

Figure 4-12 shows the risk analysis output for potential variations in the project completion date. The histogram shows the frequency each completion date was calculated during the simulation run. The simulation results are very interesting. They indicate that the deterministic project finish date, the duration calculated using the most likely duration for each activity, is December 6 and only has a 6% probability of occurring. The simulation analysis, which accounts for the variation in activity durations, indicates that the project will be completed on December 16 with an 85% probability. This illustrates the power of a Monte Carlo simulation to calculate a more realistic completion date for a project. PertMaster can be used as a planning tool to assess the probable impacts of different activity durations and configurations on the project outcome. Figure 4-13 shows that the same type of output can be generated for the individual activities in the CPM network.

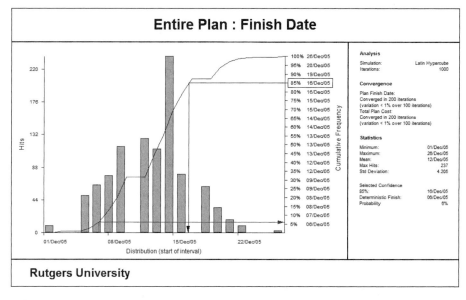

FIGURE 4-12 Total Project Duration Risk Analysis

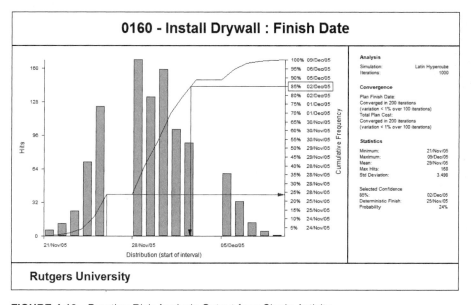

FIGURE 4-13 Duration Risk Analysis Output for a Single Activity

The PertMaster program allows for activity costs to be input as a statistical distribution. This allows the cost risk to be analyzed for a project, and also provides output that allows for a joint analysis of a project's duration and cost risks. Figure 4-14 shows the task detail window where activity costs can be entered. For this example, we defined a resource called Cost and then input the cost data for each activities cost resource. Alternatively costs can be added to actual resources such as a carpenter or materials. This would be done when importing a cost-loaded schedule from Primavera or Microsoft Project. The output for a cost–risk analysis is similar to the analysis for a duration risk.

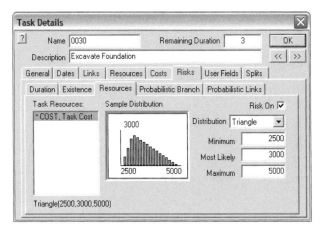

FIGURE 4-14 Entering Costs in the Task Detail Window

Figure 4-15 shows a bar chart that shows the simulated project costs. In this case the simulation shows that there will not be much variation in the total project cost, with a deterministic cost of $20,000 and an 85th percentile cost of $20,490. This analysis would indicate to a planner that changes in the project duration are to be expected, but that costs will not vary a great deal. Figure 4-16 shows a scatter plot generated by PertMaster that plots the calculated cost versus calculated duration for each simulation run. This allows a schedule analyst to view the possible interactions between cost and duration.

PertMaster has several interesting features. One of them is the ability to add probabilistic links to the network. A good example of its use is shown in the example project. An early activity calls for excavation of the footings. Perhaps we know that there is a slight chance of encountering a soil contaminated with hazardous waste at the sight. If this did occur it might greatly lengthen the project duration. To incorporate this concern in the PertMaster schedule we add an activity "Clean up hazardous waste" that will occur between excavating the foundations and pouring the foundations. However, based on our knowledge of the area, we make the preceding link a probabilistic link with only a 10% chance of occurring. This is shown in Figure 4-17. This means that approximately 1 out of 10 simulation runs will include the additional activity and we will model the potential for a significant lengthening of the project.

Clearly, PertMaster is a valuable program for studying the possible outcomes of a project, and for considering the probability of a worst-case scenario. It is further enhanced by its capability to exchange data with the most popular scheduling programs. If the activity durations for a project are uncertain, PertMaster can provide a useful way of assessing the degree of uncertainty. It is a particularly useful tool for large projects in which considerable uncertainty may exist about the project conditions.

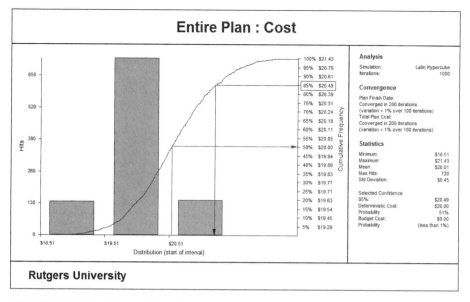

FIGURE 4-15 Bar Chart of Simulated Project Costs

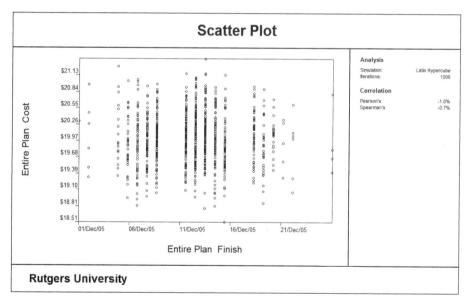

FIGURE 4-16 Scatter plot of Cost Versus Duration for Simulation Runs

Scheduling and the Computer

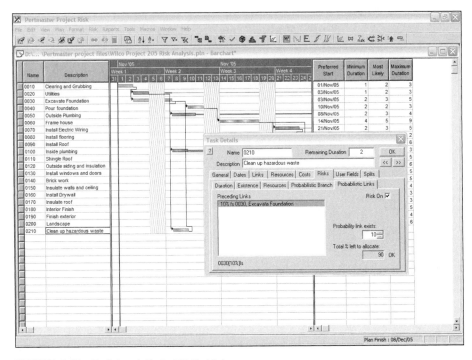

FIGURE 4-17 Defining A Probabilistic Link

■ Conclusions

Computer programs to perform CPM scheduling have long been used in the construction industry. The most commonly used programs, Primavera and Microsoft Project, offer many features for scheduling a complex project. Software like PertMaster extends the features of the major programs by allowing constructors to model the uncertainties inherent in duration and cost. The software is available that lets these programs run on a LAN so multiple users can modify and access schedule information. The major software programs now offer web-based access for program users in which schedule information can be obtained via a web browser. The program now includes features that let top management get information on a portfolio of their firm's projects. The features of these programs and the ability for users to access them from remote locations will continue to evolve. Scheduling using computers will clearly remain one of the primary IT applications in construction.

■ References

Hinze, J. 2003. *Construction Planning and Scheduling.* Upper Saddle River, NJ: Pearson Prentice Hall.

Kelleher, A.H. 2004. An Investigation of the Expanding Role of The Critical Path Method by ENR's Top 400 Contractors. Master's Thesis, Virginia Polytechnic Institute and State University.

Microsoft Corporation. 2004. Project 2003: Which Offering is Right for You? 2004. http://www.microsoft.com/office/project/howtobuy/choosing.mspx (Accessed November 3, 2005)

Newitt, J. 2004. *Construction Scheduling Principles and Practices.* Saddle River, NJ: Pearson Prentice Hall.

O'Brien, J. and Plotnick, F. 2005. *CPM in Construction Management.* New York: McGraw Hill Professional.

Primavera Systems, Inc. 2005. *Primavera Administrators Guide Version 5.0.* Bala Cynwyd, PA: Primavera Systems, Inc.

Weber, S. 2004. *Scheduling Construction Projects: Principles and Practices.* Upper Saddle River, NJ: Pearson Prentice Hall.

CHAPTER **5**

Internet-Based Solutions for Small Companies and Projects

INTRODUCTION

Many of the existing web-based applications for the construction industry have focused on information exchange and document management for large projects. Small contractors often do not have the skills or the resources to usefully implement IT solutions such as a complex web portal solution. In addition, the scope of many construction projects does not justify the investment in web portal systems. New developments in web-based software and networking technology are making it increasingly possible for construction managers to use the web inexpensively. The new IT solutions do not require extensive computer programming knowledge, and can therefore be used by a wide variety of people in a construction company.

These new technologies have the potential to greatly expand the use of IT in the construction industry and provide more firms, including small contractors with little IT experience, with the increased efficiencies possible with the use of IT. In this chapter we will discuss three relatively simple techniques that have recently emerged and can increase information and knowledge exchange on construction projects. We will discuss two web-based programs, weblog and wiki, and consider the use of peer-to-peer networking by exploring the Groove software.

◼ Weblogs

Weblogs have many potential uses in the construction industry due to their ease of use and flexibility. There has been considerable recent interest in weblogs for their capability to instantly transmit news and opinions. Some journalists publishing weblogs of political opinion have found wide audiences. In some industries, weblogs have been used as a collaborative tool for which employees share and comment on ideas.

Introduction

A weblog is composed of brief, frequently updated posts that are arranged chronologically (Bausch et al. 2002). No knowledge of computer programming or HTML is required to post information to a weblog and no special software is required. Only web access and a web browser are required to access a weblog and add posts (Stone 2002).

Their ease of use makes weblogs an ideal application for the construction industry, in which many people may not have extensive computer knowledge. Typically weblog posts are automatically archived. This can be an advantage over e-mails because the message is archived in a central repository that everyone can access.

The weblog also can provide a means of fostering collaboration. There are various modes of using a weblog. One way is for a single person or entity to provide all of the posts. Most weblog systems provide a feature for commenting. This allows weblog users to comment on posts that are made, which are added to the weblog. An alternative approach that can allow for enhanced collaboration is to set the weblog up to have multiple posters. This can allow a dialog to develop between the weblog users.

Weblogs for Small Contractors and Small Projects

Weblogs can be implemented at a low cost and they are easy to use. They are an ideal way to provide collaboration capabilities on small projects where the use of a web portal system may not be justified or is too expensive. Because the weblog is so easy to implement and use it can allow for the benefits of collaboration provided by much more complex systems to be used on a broad spectrum of projects.

Basic Weblog Functions

The primary feature of a weblog is the ability to post text to a web page. The user does not require any knowledge of HTML or computer languages to input the information. Typically, weblog posts are arranged chronologically and are given a time and date stamp. In other words, the weblog provides a way to easily create web content, and send it to the WWW without requiring significant technical knowledge other than the capability of using a web browser.

Weblogs can also be used to foster collaboration. Commenting is a feature that many weblog systems provide. Someone viewing the weblog can attach comments to the post, which are viewable by all weblog users. This feature allows weblogs to be used for collaboration and as a forum on which ideas can be exchanged.

In a weblog, posts are not discarded. Instead an archive of older posts is created, and made accessible to the user. This archive is desirable for construction applications where the weblog can serve as a record of project incidents and milestones.

Evolving Weblog Capabilities

As the weblog has evolved over time, various providers and software have emerged that provide enhanced capabilities. Primary among these capabilities is the ability to attach files to weblog posts so they can be shared with other weblog users. Another useful feature in a construction context is the ability to post photos to a weblog. A natural use of this feature for construction is to provide and distribute progress photos. Additionally, photographs of construction problems can be posted to a weblog to be rapidly sent to project team members.

Categorize Posts

Many weblogs provide the capability to categorize the posts made within a weblog. You have the capability of assigning a key word to a post such as "asphalt" or "concrete." It is then possible for the weblog to display posts related only to a particular category.

Security Features

A primary concern in the construction industry is to provide security for sensitive project data and the proprietary knowledge of the construction firm. The original weblogs that were developed did not have any security features. The weblogs were accessible by anyone who was able to locate the URL for the web page. Weblogs are now available that provide security.

Audio and Video

Many weblog systems now have the ability to include audio and video clips in weblog posts. This can provide a variety of possibilities for exchanging and recording project information. The ability to include audio in weblogs has recently become easier through the emergence of podcasting. Anyone with a computer, the free iTunes software, and a microphone can produce audio files in standard formats for use on a weblog. The Blogger weblog service has the capability to send audio files as posts to a web site. This capability allows anyone to immediately post important audio announcements to a construction web site. Potentially, this is an easier way for busy construction people to interact using a weblog.

Available Weblog Services and Software

Weblogs can be provided in two different ways. The first is to belong to a weblog service that provides hosting for a company's weblogs. The second way is to obtain weblog software and host the weblog on the firm's computers. For a small company it would be possible to host the weblogs using a personal computer as the server.

Both methods have advantages and drawbacks. Many construction companies have serious concerns about the security of their data. In working with a construction company with significant IT capabilities it was found that they preferred to host their own weblog.

Another additional concern with hosted services is that a stable hosting provider be selected. Some companies are small and may go out of business, with the subsequent loss of project data. However, there are several well-known weblog service providers that provide a safe haven for project data.

Of the weblog services available, many offer their services for free. However, the free services offer only basic features. Typically, these free services do not provide any security features for the web site that is created. Anyone surfing the web can view the weblog web page. This lack of security typically rules out the use of free weblogs by construction companies for project management applications. However, weblog services that are provided for a fee typically provide password protection for the weblog web site.

The original weblog service is Blogger. Anyone can establish a weblog for free at the Blogger site (www.blogger.com). Blogger has several powerful features, but is limited in its construction applications because it does not offer passwords. There are many services available for a monthly fee that provides adequate security and many functions. Web links for some popular hosting services are shown at the end of the chapter. There are many available and this list is by no means exhaustive. Table 5-1 shows details about two well-known weblog services, TypePad and Radio Userland.

TABLE 5-1 Weblog Services

Name	Price	Features
TypePad	• $14.95 per month for a pro account with multiple users and unlimited number of weblogs	• offers paid accounts with many features • password protection, photo uploading, file transfer, and blog posting from mobile phones are provided • an account can be set up that allows multiple weblog authors and an unlimited number of weblogs
Radio Userland	• $39.95 • includes a year of hosting and storage of up to 40 mb	• requires a program that runs on each users personal computer • application connects to a remote community server to upload posts and files to a hosted weblog • additional storage for weblog data can be purchased if necessary • can be uploaded to any server using File Transfer Protocol (FTP) for stand alone applications • includes a newsreader, and extended content management capabilities such as the ability to upload and share text documents from a page on the weblog site. • a drawback is that it requires each person who will post to the weblog to have the Radio Userland software installed on his or her computer (Cadenhead 2004).

The best-known software tool for providing your own weblog hosting on a server is called MoveableType (Six Apart, Ltd., 2003). Moveable Type is open-source software, which means that the original source code of the program can be modified and extended. The cost of a commercial license ranges around $199.95 for 1 to 5 users to $1299.95 for 50 users. Another widely used weblog hosting software is called Greymatter. Greymatter is a shareware software package and available for free (Grey 2001).

Weblog Example

To illustrate the uses of weblogs in construction we present a description of a system of weblogs developed for managing the construction of a large high school project. These weblogs were maintained by the project's construction manager.

SilkBlogs

The weblogs were implemented using the SilkBlogs weblog service. This was a hosted service, and no software or equipment was required to implement the system. The SilkBlogs service was employed because it had good security features, allowed photographs to be shared, and provided a file-sharing capability (Silk Road Technology 2004).

System Security

In this project, various levels of security were possible using passwords to control who had access to the weblog and who could add content to the weblog. There was a central coordinator who controlled access to the weblog and issued invitations. This person had the authority to add posts to the weblogs and could remove any posts if necessary. Invited users could be given various levels of access to the weblogs. High-level users were granted the right to add posts and comments to the weblog. Low-level users were only given the right to view the weblog posts and comment on weblog posts.

System Structure

Several weblogs were used to assist in managing the construction project. These included a weblog of project progress photos, a weblog of project meeting minutes, a calendar of upcoming meetings, a weblog of project financial data, and a weblog detailing outstanding project tasks. Users entering the weblog see the table of contents page with links to the other weblogs, shown in Figure 5-1.

Adding information to the weblog

To add information to the weblog, an editing window is displayed to the user. The user can type a text message in the window or paste a photo file to the window. No knowledge of any commands is required to post the information to the weblog page. Access to the Silk-Blogs editor is web based so a content editor does not require any special software to make posts to the web site. Text information can be cut and pasted from existing documents.

Task List Weblog

Figure 5-2 shows an example of a page from the task list weblog. It illustrates a typical weblog page with a chronological listing of posts. The task list weblog contains a list of outstanding project items. Posts are listed in chronological order and are automatically given a date and time stamp. Each individual post can be expanded to show more detail. The weblog also contains a category area. A weblog post may be assigned to any category created by the user. In this case the categories represent different projects being conducted for the client.

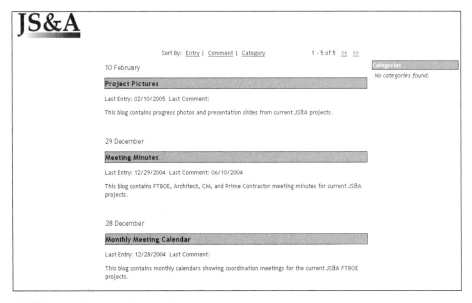

FIGURE 5-1   Weblogs Table of Contents
Courtesy of: SilkRoad Technology

Figure 5-2 also illustrates how the commenting feature can be used for collaboration. Using the commenting feature of the weblog, project staff is able to comment on the status of the tasks and exchange information. Figure 5-3 shows a weblog post expanded to show the original post and a comment about the status of the task from another project engineer.

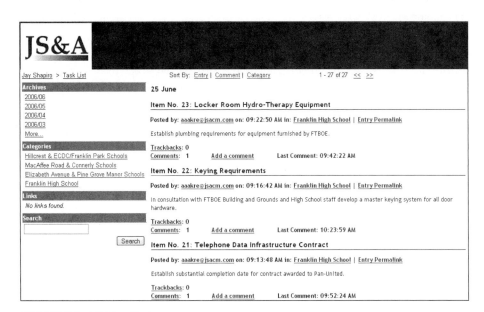

FIGURE 5-2   Weblog Posts
Courtesy of: SilkRoad Technology

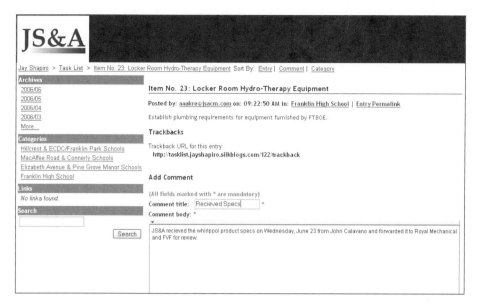

FIGURE 5-3   Weblog Post Expanded to Show Comment
Courtesy of: SilkRoad Technology

Digital Photographs

The ability to include digital photographs as a weblog has proven to be extremely useful and has become the most used weblog among both the project manager's staff and personnel from other organizations. This weblog shows project progress photos, and includes aerial photographs of the project that are taken every few months. The photos are added to the weblog simply by cutting and pasting jpeg files.

Weblog Users

Several parties both within and outside of the construction manager's organization used the weblogs. The average number of hits per day to the weblogs is 6.39. There are approximately 10 users. External organizations with access to the weblogs included the architect, subcontractors, and owner. It was found that the weblogs provide an alternative to e-mailing each participant project documents and messages. Two weblog repositories were established to share documents with project participants. A weblog that contained the monthly invoices and a weblog that contained the minutes of project meetings were included in the system. Figure 5-4 shows the weblog for project meetings in which each individual post contains a link to the project meeting for a particular date. Figure 5-5 shows an individual post from the listing of monthly invoices. A link to the invoice PDF file is provided for download.

The photographic weblog provides a way for all project participants to observe project progress and see pictures of project problem areas. Before implementation of the weblog, the construction manager had been preparing Microsoft PowerPoint presentations containing photos of project progress for the owner. The weblog allowed the owner to access the progress information directly, resulting in a time savings for the project manager. Subcontractors were also able to download the photographs to document their work.

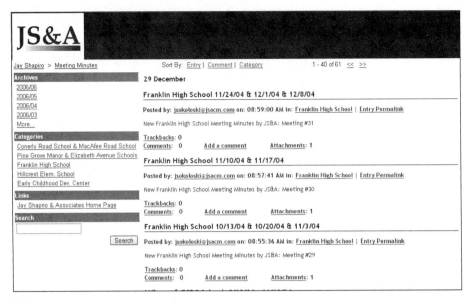

FIGURE 5-4   Weblog Used as a Repository of Project Meeting Minutes

FIGURE 5-5   Link to a Project Document Contained in a Weblog Post

Success of the Weblog Implementation

The weblogs developed provided a useful, time-saving tool for the high school project. Several elements were present that contributed to the success of this implementation. These include:

1. The attitude of the top management of the construction management firm was enthusiastic and supported the project. The construction manager's top management viewed the implementation as a way of increasing the firm's prestige by demonstrating the firm's willingness to innovate to clients.
2. The construction manager provided resources to insure that the weblogs were implemented properly. An engineer was designated to implement and maintain the weblogs.
3. The implementation of the weblog was an incremental innovation. The construction manager was already using computers extensively to manage the project, and the web-based weblogs were easily incorporated into the firm's day-to-day operations.

Considering Potential Weblog Applications

The preceding example illustrates a useful way of using a weblog for management of construction projects. However, weblogs are flexible tools, and have been adapted to many different types of uses and applications. It is useful to consider the different ways they can be used within a construction organization.

Project-Based Applications of Weblogs

Due to the project-based nature of the construction industry, there are several justifications for setting up a weblog for information and knowledge exchange for individual projects. They are:

1. Better documentation of a construction project

   - Weblogs can serve as a repository of information and an archive of project events.
   - Weblog posts are "persistent," and cannot be lost or deleted like e-mail messages.
   - Older posts are archived and stored permanently.
2. Improved collaboration between project participants

   - Participants in the construction project can exchange ideas through posts and comments on posts.
   - These exchanges are visible to all project participants.
3. Weblogs can provide a place to store the knowledge and "lessons learned" during the construction project.
4. Weblogs can potentially lead to shorter decision cycles by providing a discussion forum to rapidly address the questions and problems that arise during construction.

Topic and Function Based Applications

Weblogs can also be constructed for single topics and by function within the construction firm to provide a place to accumulate the construction firm's expert knowledge about the best ways to perform construction operations and manage the project. For example, a weblog about a type of construction, such as asphalt paving, or a weblog about a function within the firm, such as estimating, can be maintained. These weblogs can become a long-term repository for the firm's knowledge about construction operations and procedures.

Often in construction, inexperienced engineers and managers are assigned to tasks they know little about. Potentially, a weblog containing the firm's best practices could be used to provide information and more rapidly answer questions that occur in the field.

Therefore, it is possible to consider that an employee could be a member of several weblogs. For each project, the employee is working on he can participate in the project's weblog to keep abreast of the latest project information. The employee can also participate in the function-based applications to tap into the firm's knowledge about particular topics.

RSS feeds

An emerging feature of some weblog and wiki software is that they can automatically produce Real Simple Syndication (RSS) feeds. RSS feeds allow content, such as weblog posts to be shared from one web site to another web site, or to a news aggregator program. A weblog or wiki can provide a feed to its content by producing an RSS document available via a URL. An RSS document is an XML file that contains a number of discrete news items, such as entries in a weblog (Skonnard 2004). An RSS news aggregator is a program that reads RSS documents and displays new items. Most aggregators make it possible to subscribe to a feed simply by entering the URL of the RSS document. RSS feeds are becoming increasingly popular and services like Yahoo! and Google are now making it possible for users to add favorite RSS feeds to their personal pages.

RSS feeds to monitor multiple weblog memberships

The capability to employ RSS feeds may be particularly useful as a method for construction managers to track new posts from multiple weblogs. It can allow a construction engineer with many ongoing projects and membership in several weblogs to rapidly scan the latest updates and questions received from the field. A news aggregator can display lists of the titles of recently received posts from multiple-weblog web sites. A news aggregator program could serve as a place that accumulates the latest developments from all projects and topics of interest.

Wiki

Wiki is another web-based system that is becoming increasingly popular. They can provide the capability of building a complex web site without any knowledge of HTML.

Definition of Wiki

Wiki is another emerging type of web-based system that can provide an easy-to-use environment for information management, knowledge management, and collaboration. A wiki (a term derived from a Hawaiian word for fast) is defined as a freely expandable collection of interlinked web "pages." Wiki can function as both a hypertext system for storing and modifying information and a database where each page is easily editable with a web browser. A wiki allows everyday users to create and edit any page in a web site. Web pages can be edited using only a browser. A wiki user is not required to have any special software on their computer to modify a wiki.

Wiki software comes in several variations, and most of the wiki software is open source. The software is loaded on a web server. Most wiki software packages require a web server running PHP and MySQL. Wiki hosting services are also available. One example is the SeedWiki site that allows users to establish a hosted wiki for free.

Wiki users can input text on existing web pages, and can easily create new linked pages with knowledge of only a few commands. This promotes content composition by users without knowledge of HTML or web programming. One of the drawbacks of wiki is that most implementations are text only. Typically, graphics and photographs cannot be easily incorporated in wiki pages.

Wiki function in a simple manner. A wiki allows a user to edit any page within the wiki web site or to create new pages using only a web browser. Users can input or paste text into any wiki page. Link creation between different pages is automatically achieved by typing two capitalized words with no space between the words. The wiki software automatically sets up the new linked page. Wiki software typically provides built-in search capabilities to search for any word contained in the database.

In construction projects, wiki can be used in several ways to transfer both knowledge and information. These include:

- Collections of the construction firm's written documentation.
- Collaborative FAQs (Frequently Asked Questions) where construction experts respond to questions about construction techniques from personnel in the field. The wiki can provide a way for experts at other locations to interact with less experienced personnel in the field.
- A way to capture new knowledge about construction procedures. "Lessons learned" during a construction project can be transmitted to other project team members and archived on the wiki site. Experts can be encouraged to record their thoughts about best practices.

Like weblog software, wiki has the potential to provide a collaborative environment that is easy to use and low in cost. Using handheld computers, construction field personnel can enter questions, update progress status, and search for knowledge about the best ways to conduct a construction operation. Construction managers and experts can monitor the inputs from the field and respond to inquiries. A wiki web site can potentially serve as a dynamic archive of the issues that occur during construction.

Wikipedia

An excellent example of a general-purpose use of a wiki is the Wikipedia (www.wikipedia.com). This is a wiki site where an encyclopedia has been developed through the contributions of site users. Anyone with a web browser can add new information to existing topics and create new topics for inclusion in the encyclopedia. This wiki provides an example of how information and knowledge can be gathered and reused. The site has grown into a massive encyclopedia that illustrates how a wiki can be used to aggregate knowledge. It also illustrates how a wiki in its basic form has little security and can be accessed by anyone. In a construction context it would be necessary to implement a wiki with security features to prevent access by unauthorized personnel.

Wiki Construction Example

A prototype wiki has been developed to demonstrate the capabilities of a wiki web site for construction project knowledge management. The software employed is called QuickiWiki (Leuf & Cunningham 2000). The software requires a Perl interpreter to be operated as a stand-alone program on a computer. Alternatively it can be set up to run on a web server so it is accessible on an intranet or the Internet as a public wiki. The prototype presents a wiki for a highway construction project. It demonstrates several ways a wiki can be used for knowledge management and information exchange. Figure 5-6 shows the opening page of the wiki. The page contains internal links to other pages within the wiki with information about highway construction techniques including paving and traffic signals and an external link to some construction specifications. A search function is also included on the opening page. The wiki software is capable of searching for text within all of the different pages in the wiki. A search produces a page listing all of the pages that contain the search text.

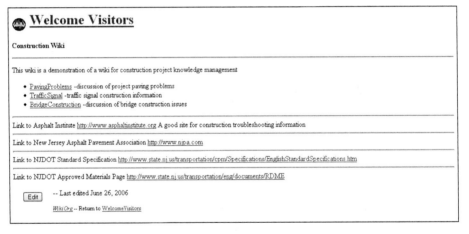

FIGURE 5-6 Wiki First Page
Courtesy of: Quiki Wiki

Figure 5-7 shows the edit window for the opening page. It is in this edit window that new information can be added to the page. The edit page contains some simple wiki formatting commands such as triple apostrophes to highlight text and the asterisk to make bulleted text. In the basic QuikiWiki software it is possible for any user to modify the content of any page by clicking on the "edit" button that is displayed at the bottom of every page. However it is possible to revise the software to allow pages to be locked so only a system administrator can modify them. Additionally, most wiki software has the capability to provide passwords that restrict access to the wiki site.

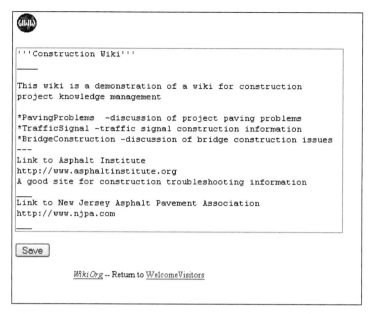

```
'''Construction Wiki'''

────
This wiki is a demonstration of a wiki for construction
project knowledge management

*PavingProblems  -discussion of project paving problems
*TrafficSignal -traffic signal construction information
*BridgeConstruction -discussion of bridge construction issues
---
Link to Asphalt Institute
http://www.asphaltinstitute.org
A good site for construction troubleshooting information

────
Link to New Jersey Asphalt Pavement Association
http://www.njpa.com
────
```

[Save]

Wiki Org -- Return to WelcomeVisitors

FIGURE 5-7 Wiki Edit Window
Courtesy of: Quiki Wiki

Other pages within the Wiki can be used as a place for constructors to ask questions about construction problems. It illustrates the potential collaborative capabilities of the wiki by allowing for people to discuss construction problems and create "lessons learned" that everyone in the firm can see. Wiki pages can also contain links to internal and external web pages. Users with useful links can add them to wiki pages to be shared.

■ Peer-To-Peer Networking For Construction Project Management

Peer-to-peer networking can provide a method for team members on small construction projects to link electronically without the computer infrastructure required for more complex applications. Peer-to-peer networks are a way of exchanging data without the need for a central server or application service provider.

Introduction

Systems employing peer-to-peer networking seem to have great potential for application for both knowledge and information management for construction projects. Peer-to-peer networking is a type of networking in which a group of computers communicate directly with each other, rather than through a central server. This may be attractive for small construction companies because they can set up a network without the need for a client-server system.

Peer-to-Peer versus Client-Server Architecture

The systems we have discussed so far in this book have employed a client-server architecture. The programs we have talked about require servers dedicated to serving other computers. Alternatively a peer-to-peer network is type of network in which each workstation has equivalent capabilities and responsibilities. The use of peer-to-peer networks involves a trade-off when compared to client-server architecture. With a client-server arrangement you only need a web browser to access the web-based software. However, you must maintain a computer as a server, or subscribe to a service that maintains the infrastructure for you (such as a web portal service). With a peer-to-peer network the user interface runs outside of a web browser. Typically you must purchase and install the peer-to-peer software on each computer in the network. The advantage is that there is no server infrastructure that needs to be maintained. You can have a network without any of the associated equipment costs. This may be highly desirable for a small construction company or for a small project where a small network is maintained.

Typical characteristics of peer-to-peer systems include the ability for computers in the system to act as both clients and servers, the software is easy to use and well integrated, the system includes tools to support users wanting to create content and the system provides connections with other users.

Groove

Groove is a groupware software that operates using peer-to-peer networking. Groove allows users to easily set up small networks by creating a shared workspace to collaborate with other Groove users. A Groove user may establish multiple workspaces. The Groove networks are easy to set up, requiring no programming or IT expertise. A Groove network can be established by users without any networking experience and can be done without the intervention of an IT department, ISP, or web hosting service. To join, all that is required is for each user to be a Groove member (i.e., to have the Groove software installed on their computer). The cost of Groove per computer is $69 for the standard edition and $149 for the professional edition (Groove Networks 2005).

Groupware

Groove fulfills the functions of groupware that was typically only available to users able to invest significantly in client-server infrastructure. Lotus Notes is probably the best-known groupware. Groupware is software that enables a group of users to collaborate on a project by means of network communications. There are several barriers to implementing Notes in the construction environment. Notes cost several hundred dollars per user, not including the expense of server computers. The Notes program also requires intense support by IT professionals. Certainly the web portals discussed in the previous section including applications such as Buzzsaw and Constructware can be categorized as construction-oriented groupware. However, due to their cost and complexity, their application has been mainly limited to large projects.

Using Groove

Establishing a workspace sets up a Groove network. Users may select various standard Groove Tools that may be included in the workspace. Several Collaborative tools are available for inclusion in a Groove Workspace (Pitzer 2002).

The available standard tools include:

1. File sharing—users in a Groove network are able to store and share files. Word and PowerPoint files can be co-edited.
2. Custom forms can be created.
3. Pictures can be shared.
4. A calendar can be maintained for members of the workspace.
5. Threaded discussions can be conducted.
6. Documents can be reviewed.
7. Instant messages can be sent. Voice messages can also be sent to workspace members.
8. Meeting minutes can be recorded and action items assigned.
9. Contacts can be shared between team members.

Groove Construction Tools

Additional add-on tools can be purchased. Relating to construction, tools are available that allow for collaborative viewing of CAD files from within a group workspace, and software that allows a Microsoft Project schedule to be imported and modified. Figure 5-8 shows a Groove Workspace with the file-sharing tool displayed. Along the bottom of the screen is the text-messaging window.

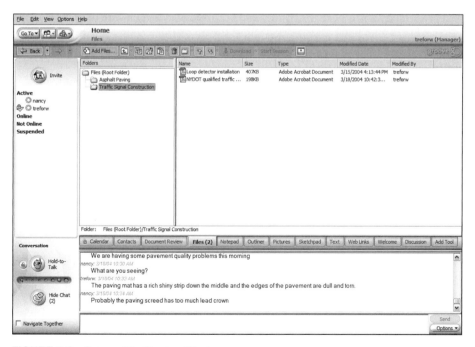

FIGURE 5-8   Groove File Sharing Window
Copyright: 2000–2005, Groove Network, All Rights

Internet-Based Solutions for Small Companies and Projects

Using the Groove Workspace Project Edition it is possible to create schedules in a Groove Workspace. The TeamDirection project tool allows Microsoft Project files to be imported to a Groove Workspace and also exported. It is also possible to keep the Groove workspace synchronized with the Microsoft Project file. The synchronization option allows a project that has been created within Microsoft Project to be loaded into a Groove workspace to allow for collaboration between team members, and still allow for the schedule to be updated periodically in Microsoft Project.

A CADViewer application is available as an add-in tool for Groove. The tool allows 2D CAD drawings to be viewed alone by a Groove user or to be viewed in real time with other workspace members.

Other Useful Peer-to-Peer Products

Several other pmail users often encounter difficulties sending large files as attachments to e-mails. Some e-mail systems limit the size of e-mail attachments. Peer-to-peer file systems can be used as an inexpensive method of sharing large files.

Groove is the most powerful, commercially available, peer-to-peer networking software available. However other peer-to-peer networking tools are available that can be useful in a construction context. These tools have a more limited repertoire of capabilities than Groove but can still be useful as a way of exchanging files or rapidly contacting a project participant via instant messaging. LapLink sells a file-sharing program called ShareDirect that allows a user to share files and folders with other ShareDirect users across the Internet. ShareDirect also has instant messaging features. Share direct pricing varies based on the amount of data that is sent per month and varies from $39.95 per year to $249.95 per year for a corporate account.

Qnext is another peer-to-peer program that is currently available free of charge. The program allows users who have downloaded the software to their computer to join groups and exchange files, participate in instant messaging and online text meetings, and share photographs. Figure 5-9 shows the QNext file sharing window. In the example two QNext users are exchanging files of bid tabulations. Figure 5-10 shows the online instant messaging features of QNext that could be used to exchange information in real time between construction firm members.

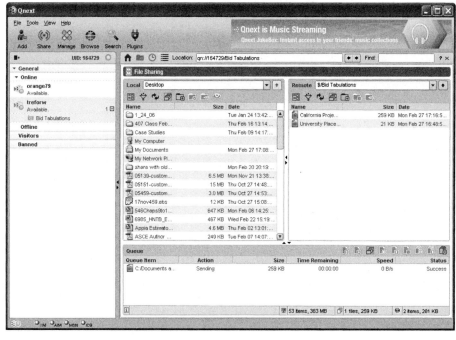

FIGURE 5-9   Peer-to-Peer File Sharing Using QNext

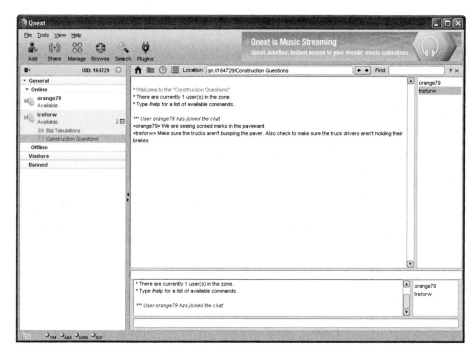

FIGURE 5-10   A QNext Online Chat Session

Internet-Based Solutions for Small Companies and Projects

■ Web Sites

The following table provides links to the weblog and peer-to-peer networking software discussed in this chapter.

Name	Service	Website
Blogger	Weblog Service	www.blogger.com
LiveJournal	Weblog Service	www.livejournal.com
TypePad	Weblog Service	www.typepad.com
RadioUserland	Weblog Service	radio.userland.com
SilkBlog	Weblog Service	www.silkware.com
MoveableType	Weblog Software	www.moveabletype.org
Greymatter	Weblog Software	www.noahgrey.com/greysoft
Groove	Peer to Peer	www.groove.net
ShareDirect	Peer to Peer	www.laplink.com/products/sharedirect/overview.asp
QNext	Peer to Peer	www.qnext.com

■ References

Bausch, P., Haughey, M., and Houlihan, M. 2002. *We Blog: Publishing Online with Weblogs*. Indianapolis: Wiley Publishing, Inc.

Cadenhead, R. 2004. *Radio UserLand Kick Start*. Indianapolis: Sams Publishing.

Grey, N. 2001. *Greymatter-The Opensource Weblog/Journal* http://www.noahgrey.com/greysoft (Accessed 14 April 2004).

Groove Networks. 2004. *Groove Workspace Benefits*. http://www.groove.net/default.cfm?pagename=WSBenefits (Accessed 1 February 2004).

Leuf, B. and Cunningham, W. 2000. *The Wiki Way: Quick Collaboration on the Web*. Boston, MA: Addison-Wesley.

Pritzer, B. 2002. Special Edition Using Groove 2.0. Indianapolis: Que Publishing.

Six Apart Ltd. 2003. Moveable Type. http://www.moveabletype.com (Accessed 15 March 2004).

Skonnard, A. 2004. The XML Files: All About Blogs and RSS, *MSDN Magazine*. http://msdn.microsoft.com/msdnmag/issues/04/04/XMLFiles/default.aspx (Accessed 9 April 2004).

Stone, B. 2002. *Blogging: Genius Strategies for Instant Web Content*. Indianapolis: New Riders Publishing.

CHAPTER 6

Construction Web Portals— For Large and Complex Projects

INTRODUCTION

On complex projects, thousands of documents may be exchanged between the project participants.

Methods like a weblog or wiki are not sophisticated enough to handle the volume of paperwork. A method

to control, route, and modify project documents between the owner, designer, contractor, and subcon-

tractors is required. This chapter discusses web portals, which have become a popular way of enhancing

document exchange and collaboration on large projects.

■ What is a Web Portal?

A web portal (sometimes called a project extranet) is a web-based service that allows for collaboration and document exchange on construction projects. A web portal provides a web page for a construction project that allows users to access project documents and collaborate with each other through the exchange of messages.

The basic idea of a construction web portal is to apply web-based systems and appropriate construction project management techniques to facilitate online collaboration through better communication and workflow management. The web portal thus provides efficient information exchange to assist in successfully delivering a project.

Potential Benefits of Using a Web Portal

The primary reason for using a web portal is to enhance communication between the project participants. In construction projects poor communications is believed to impact projects significantly by causing delays and inefficiencies. A web portal allows key project information to be available to all participants and, if used properly, provides project participants with timely information about the project status.

A benefit that is possible when employing a web portal system is to reduce the number of paper plans and documents that are required. Project participants can access project documents electronically through the web-portal and do not need to print documents.

A Central Point for Information Exchange

A construction project web-portal provides a central point where all project participants can interact. The many players in a construction contract including the owner, designer, prime contractor, construction manager, and subcontractors can all access the web portal. Without a web portal, communications can be chaotic and fragmented with project participants communicating and exchanging documents by telephone, e-mail, mail, and fax. The web portal provides a repository of all project documents.

Types of Documents Exchanged

Many different types of documents are exchanged on a typical construction project. On a large project, thousands of documents are generated. Web portals allow documents such as CAD files containing the project plans, change orders, meeting minutes, requests for information, inspection reports, and shop drawings to be exchanged.

A major feature of web portal systems is to provide document-tracking capabilities. Typically web portals can track who has viewed and modified documents. This can be particularly important on a complex project for which it may be necessary to track hundreds of change orders and requests for information simultaneously. The status of changes and questions can be quickly ascertained so the project schedule is not impacted.

■ Integration of Design and Construction

A major benefit of web portals is the enhanced integration of the design and construction process. A web portal for a project can be initialized during the design phase of the project, and all design information is available through the web portal. During construction, required design changes and questions about the design can be effectively tracked through the web portal. Communication between the designer and contractor is enhanced and are faster than traditional paper-based methods.

■ Web Portal Service Providers

Web portal software can be purchased or leased and installed on a construction company's own server. However, many contractors access web portals as an online service provided by the web portal company. In this way contractors are able to access the sophisticated web portal software over the web without needing to invest in significant computer infrastructure. It should also be noted that a large contractor might need to subscribe to several web portals simultaneously because contract specifications may require them to use a particular web portal system on a project. Because of the many perceived benefits of using web portals on complex projects, owners now often require the use of a particular web portal system on a project.

There are many web portal service providers. These include Primavera, Prolog, Constructware, and Buzzsaw. Companies that are well known in the construction industry provide some of the web services. Expedition is provided by Primavera, the authors of the Primavera Project Planner scheduling software. AutoDesk, the developers of AutoCAD, provides the Buzzsaw service. A table of web links to some web portal service providers is given at the end of this chapter. Increasingly, web portal service providers are providing links to other software. For example, the Prolog portal provides integration with Microsoft Word, Microsoft Project, Primavera Project Planner, and Bentley ProjectWise (Meridian Systems, 2005a).

■ Web Portals and Mobile Computing

The capability has also emerged to collect data using a mobile device, connect to a computer, and upload the information to a web portal service. Prolog Pocket allows information such as daily journals, inspections and Tests, material inventories, punch lists, and safety notices to be collected. This capability can increase the speed with which data is entered in the system, and reduce field personnel's paper work (Meridian Systems, 2005b).

■ Cost of Web Portal Software

The cost of using a web portal service can be extensive, and varies based on the number of system users. The cost may be several thousand dollars a month for a web-based service. The cost of a web portal system and the expense of training personnel limits the use of web portals to larger-sized projects and firms.

■ A Web Portal Example: Constructware

To illustrate some of the capabilities of web portal systems some examples using Constructware are presented (Emerging Solutions, Inc., 2005). Constructware is a web service that provides a web-portal for construction project management. Project participants are able to access and exchange project data and information via the web portal. Constructware offers a broad range of features that are useful for design, bidding, management of the construction process, management of subcontractors and suppliers, project cost control, and project risk management. The Constructware system has the capability of aggregating cost information from multiple projects so users at the top level of a construction company can gain a clearer picture of the firm's performance.

These features include the ability to integrate design with construction by allowing for the exchange of CAD design files between project participants. Constructware provides many features that allow the broad range of paperwork that is generated during a construction project to be stored in a central repository. It allows documents to be rapidly sent to project participants, and it provides rapid means for generating a multitude of different types of documents such as change orders and requests for information.

Some Basic Features of Constructware

Figure 6-1 shows the opening view a top-level manager at a construction company would receive after logging into the Constructware web site. The opening page lists all new messages and documents received. It should be noted that the messages are input to the Constructware system by other project team members. Messages can be received from project participants and via email from people not members of the project team. They are all aggregated in the messages window. On the left of the opening screen the various Constructware modules that appear as file folder icons can be accessed.

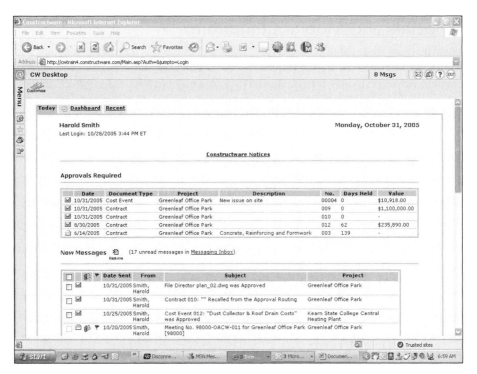

FIGURE 6-1   Constructware Today Page

Copyright: 1994–2005, Emerging Solutions, Inc., d/b/a Constructware, All Rights Reserved

The Constructware web portal can provide information about an entire portfolio of projects that a firm is undertaking simultaneously. It can be observed that documents have been received from two different projects. The icons on the left allow the user to access the various Constructware functions including the file director that allows for the uploading, downloading, and viewing of CAD documents and the document manager that allows for the exchange of other types of files and correspondence.

Dashboard

A tab at the top left of the opening screen is called Dashboard. If the user clicks on this screen a page with several user selectable charts is displayed. The purpose of the dashboard is to give managers a rapid view of the project situation and help to identify problems in their early stages. The dashboard is customizable for top executives, project managers, financial analysis, or any team member. Various performance indicators related to cost, schedule, discovery of outstanding items, and project portfolio management.

It is possible to double-click on a dashboard and receive more detailed information. For example in Figure 6-2, the top-left dashboard graphic titled "Submittal Item BIC" can be clicked on for more information. BIC stands for Ball in Court and indicates who is responsible for submittal items. If a large number are the responsibility of the user's firm it may indicate a problem. Figure 6-3 shows that after double clicking on the "Submittal item BIC" graphic, a listing of items that are outstanding and who is responsible is displayed. Each individual item can be opened from this window.

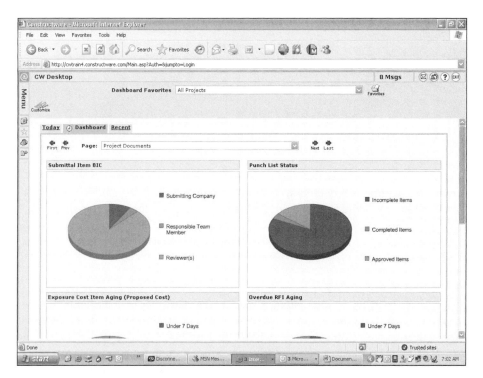

FIGURE 6-2 Dashboard View

Document Management

There are various types of documents that are handled by the Constructware system. It is possible to upload various types of files from a user's computer to the Constructware system so that these files can be shared with other project participants. The types of files that can be uploaded include CAD, PDF, word processing, faxes, scanned documents, pictures, spreadsheets, and other types of documents.

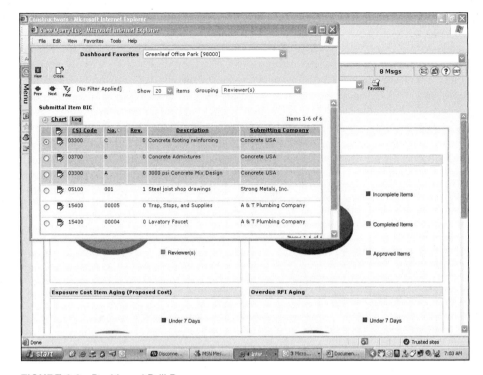

FIGURE 6-3 Dashboard Drill Down
Copyright: 1994–2005, Emerging Solutions, Inc., d/b/a Constructware, All Rights Reserved

Constructware maintains a document log for each project. This log can be viewed by selecting the document module from the document management folder. The log shows all of the files that have been uploaded to the Constructware system. The log can contain various types of documents including pictures and sketches. These files from the document log can either be downloaded for viewing or viewed as online documents by using the viewer program built into Constructware.

An interesting feature of Constructware is that some types of documents use standard templates that are stored in the system. To generate these types of documents the user accesses a simple form to fill in the details of the document. Documents that can be generated in this way include Architect Supplemental Information forms, standard letters, faxes and memos, daily reports, design reviews, issues, meetings, punch lists, requests for information, submittals, and transmittals. Clicking on their icon in the Document Management folder accesses all these modules.

For example, details can be input for a meeting. Figure 6-4 shows the meeting minute template. The meeting minutes can be automatically numbered for retrieval by the system. Type of meeting can be designated from a pull-down list such as conference call or progress meeting. A form is available to enter meeting items. There is a window to type in discussion details and the capability to associate responsibility for the meeting minute item to project participants. Figure 6-5 shows the meeting minute detail template. Meeting minutes are stored in the Constructware system and available to all participants. Tasks generated during the meeting can be tracked until completion. This is superior to generating paper or e-mail meeting minutes that could be lost or discarded.

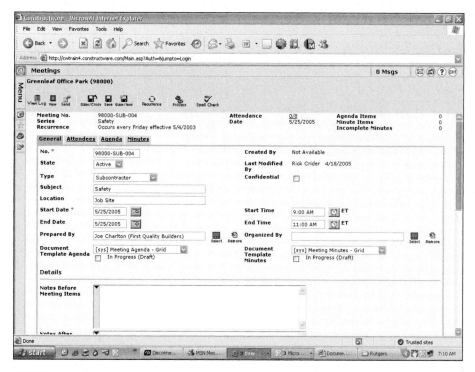

FIGURE 6-4 Meeting Setup Page

Copyright: 1994–2005, Emerging Solutions, Inc., d/b/a Constructware, All Rights Reserved

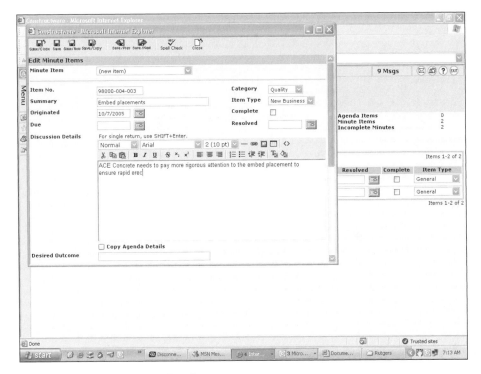

FIGURE 6-5 Meeting Minute Detail

Copyright: 1994–2005, Emerging Solutions, Inc., d/b/a Constructware, All Rights Reserved

Sending Documents and Messages to Other Project Participants

The correspondence module provides a method of creating documents that can be sent to other Constructware users using the Constructware system or sent to other individuals not members of a Constructware project team using e-mail and/or fax. A correspondence log is provided to track all types of documents. It lists all correspondence sent and received by a project team member. The log provides a list of sent and received documents. An item can be selected from the log and viewed.

The correspondence module can also be used to generate correspondence to other project participants. Figure 6-6 shows the form that can be used to send a message. It allows a user to select from a list of project participants or groups of participants to receive the message. In addition, it is easy to link documents to the message and to link the message to a particular CSI code used in the project. The correspondence module of Constructware illustrates one of the most powerful features of a web portal system. That is the capability to rapidly send messages and documents to members of the project team, speeding the flow of information on the project.

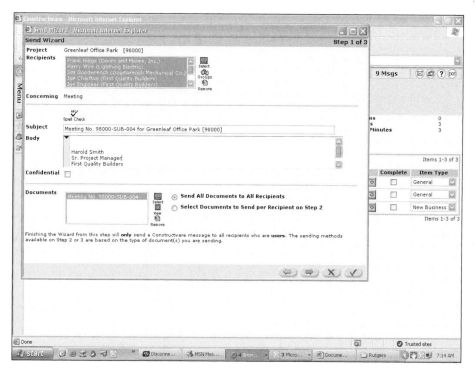

FIGURE 6-6 Sending a Document

Copyright: 1994–2005, Emerging Solutions, Inc., d/b/a Constructware, All Rights Reserved

CAD Drawings

CAD drawings are handled in the File Director module of Constructware. Figure 6-7 shows a listing of the architectural drawings for a project in DWG format. A user can either download the drawings to be viewed later in a CAD program or can use the built-in document viewer to view the file over the web without downloading it to their own computer. The document viewer is the small blue icon shown in the figure under the open column. Figure 6-8 shows an architectural plan that has been selected for viewing. Users may zoom in on different parts of the drawing to see details.

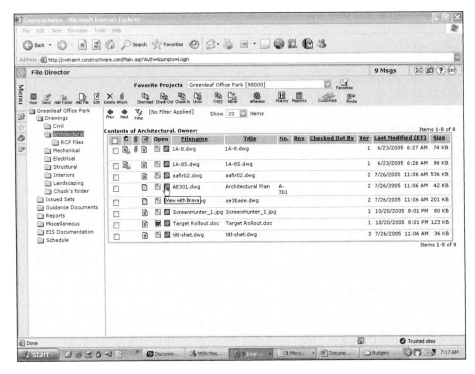

FIGURE 6-7 File Director Log

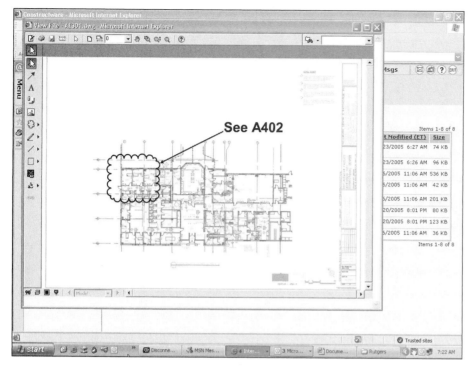

FIGURE 6-8 File View with Markup

Document Tracking and the Integration of Design and Construction

Constructware stores all documents in an electronic format and provides capabilities to track and route documents. The use of Constructware can be initiated at the early stages of a project. It is recommended that web portals be initiated early in the project so that the designer can place the CAD drawings in the web portal as early as possible so that they can be available for discussion by all project participants earlier in the project cycle. The web portal can be used as a way of achieving integration between design and construction by providing an easier way for designers to interact with construction personnel. During construction, the link between the designers and constructers can allow for more rapid solutions to design problems that are encountered during the construction phase of the project.

Constructware allows documents to be routed to project participants and tracked. The Constructware system maintains a history of who has viewed and modified documents. The routing feature allows documents to be routed for approval and checking. Design documents found in the file manager, cost documents such as change orders and cost items and issues can all be routed for review.

Constructware has an Approval Documents module within the Personal Organizer. This feature allows a document such as a change order document to be developed and routed. It is possible to attach other documents, like pictures indicating a design change, to the change order template. The user can then send the change order to the appropriate project participants who must approve the document. The Constructware system maintains a routing history tab that displays a listing of who has viewed the document and what action they have taken.

Figure 6-9 shows a change order document that has been generated in the Constructware system.

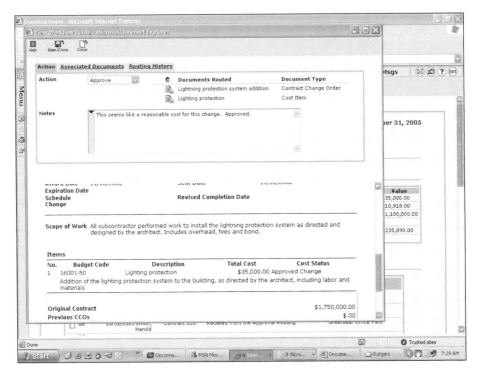

FIGURE 6-9   Change Order Approval
Copyright: 1994–2005, Emerging Solutions, Inc., d/b/a Constructware, All Rights Reserved

Financial Management and Cost Control

Constructware has many features to aid in the financial management of a construction project. The Cost Management section of Constructware has several modules that allow cost items to be defined, budgets to be developed, and the change order process to be managed. Definition of the cost items is important because they form the basic building blocks that are linked to budget and change order documents. Figure 6-10 illustrates where project cost items can be defined. Figure 6-11 shows a listing of cost items that summarize the status of each item. Budget items can be entered for a project using a standard template. Budget codes can be defined for the cost item, and estimated costs input. As the project proceeds, proposed, submitted approved, committed, and actual costs can all be entered for the item. The cost item can also be linked to standard CSI codes. Constructware can generate a budget report sorted by defined cost codes. This report allows for comparisons of budgeted versus actual costs. Figure 6-12 shows a budget screen that displays an estimate of the cost to complete a project. It allows a construction manager to quickly compare budgeted construction costs with actual expenditures, and to identify construction activities that are experiencing problems. Constructware also contains features for both preparing a bid, and receiving quotations from subcontractors. The system allows bids from subcontractors to be tracked and managed.

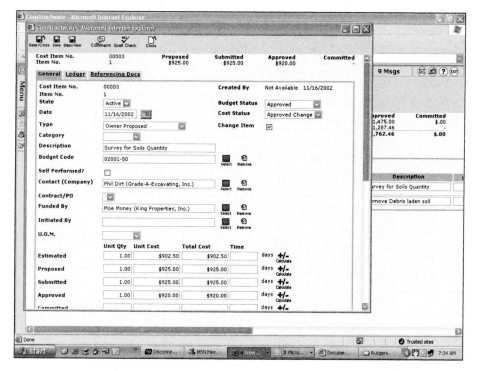

FIGURE 6-10   Cost Item on a Cost Event

Copyright: 1994–2005, Emerging Solutions, Inc., d/b/a Constructware, All Rights Reserved

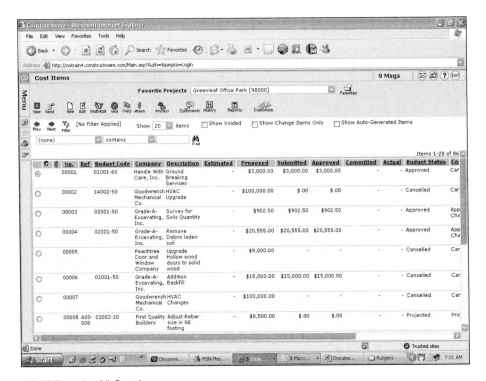

FIGURE 6-11 All Cost Items

Copyright: 1994–2005, Emerging Solutions, Inc., d/b/a Constructware, All Rights Reserved

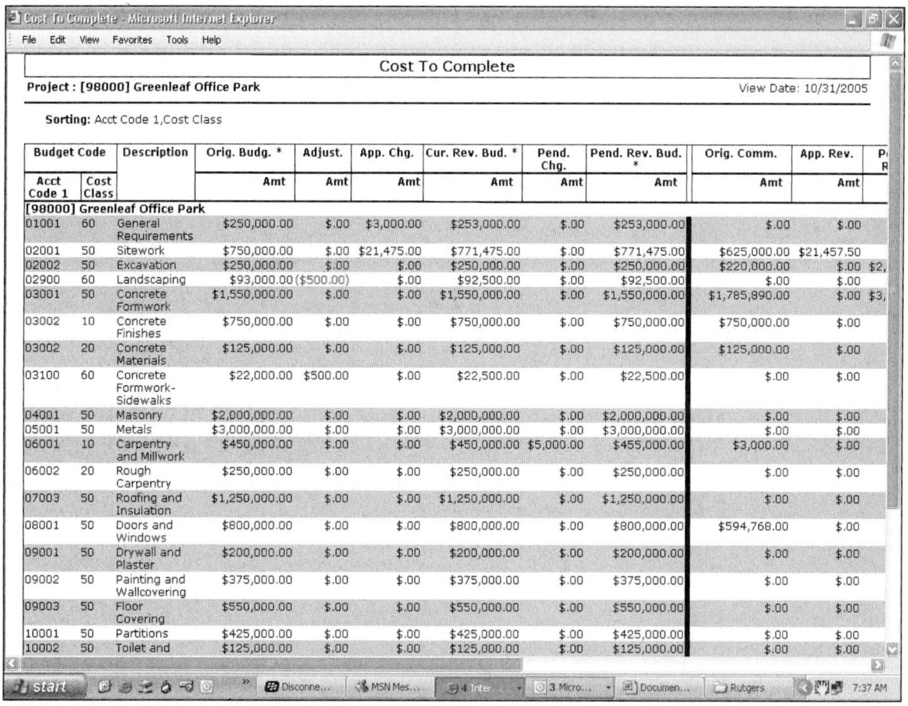

Cost To Complete

Project : [98000] Greenleaf Office Park

View Date: 10/31/2005

Sorting: Acct Code 1,Cost Class

Budget Code		Description	Orig. Budg. *	Adjust.	App. Chg.	Cur. Rev. Bud. *	Pend. Chg.	Pend. Rev. Bud. *	Orig. Comm.	App. Rev.	P R
Acct Code 1	Cost Class		Amt	Amt	Amt	Amt	Amt	Amt	Amt	Amt	
[98000] Greenleaf Office Park											
01001	60	General Requirements	$250,000.00	$.00	$3,000.00	$253,000.00	$.00	$253,000.00	$.00	$.00	
02001	50	Sitework	$750,000.00	$.00	$21,475.00	$771,475.00	$.00	$771,475.00	$625,000.00	$21,457.50	
02002	50	Excavation	$250,000.00	$.00	$.00	$250,000.00	$.00	$250,000.00	$220,000.00	$.00	$2,
02900	60	Landscaping	$93,000.00	($500.00)	$.00	$92,500.00	$.00	$92,500.00	$.00	$.00	
03001	50	Concrete Formwork	$1,550,000.00	$.00	$.00	$1,550,000.00	$.00	$1,550,000.00	$1,785,890.00	$.00	$3,
03002	10	Concrete Finishes	$750,000.00	$.00	$.00	$750,000.00	$.00	$750,000.00	$750,000.00	$.00	
03002	20	Concrete Materials	$125,000.00	$.00	$.00	$125,000.00	$.00	$125,000.00	$125,000.00	$.00	
03100	60	Concrete Formwork-Sidewalks	$22,000.00	$500.00	$.00	$22,500.00	$.00	$22,500.00	$.00	$.00	
04001	50	Masonry	$2,000,000.00	$.00	$.00	$2,000,000.00	$.00	$2,000,000.00	$.00	$.00	
05001	50	Metals	$3,000,000.00	$.00	$.00	$3,000,000.00	$.00	$3,000,000.00	$.00	$.00	
06001	10	Carpentry and Millwork	$450,000.00	$.00	$.00	$450,000.00	$5,000.00	$455,000.00	$3,000.00	$.00	
06002	20	Rough Carpentry	$250,000.00	$.00	$.00	$250,000.00	$.00	$250,000.00	$.00	$.00	
07003	50	Roofing and Insulation	$1,250,000.00	$.00	$.00	$1,250,000.00	$.00	$1,250,000.00	$.00	$.00	
08001	50	Doors and Windows	$800,000.00	$.00	$.00	$800,000.00	$.00	$800,000.00	$594,768.00	$.00	
09001	50	Drywall and Plaster	$200,000.00	$.00	$.00	$200,000.00	$.00	$200,000.00	$.00	$.00	
09002	50	Painting and Wallcovering	$375,000.00	$.00	$.00	$375,000.00	$.00	$375,000.00	$.00	$.00	
09003	50	Floor Covering	$550,000.00	$.00	$.00	$550,000.00	$.00	$550,000.00	$.00	$.00	
10001	50	Partitions	$425,000.00	$.00	$.00	$425,000.00	$.00	$425,000.00	$.00	$.00	
10002	50	Toilet and	$125,000.00	$.00	$.00	$125,000.00	$.00	$125,000.00	$.00	$.00	

FIGURE 6-12 Cost to Complete
Copyright: 1994–2005, Emerging Solutions, Inc., d/b/a Constructware, All Rights Reserved

■ Projectwise: A Web Portal to Manage Design Information

ProjectWise is software that has a different emphasis than many web portal systems. We have seen how the Constructware software can assist in collaboration during the construction phase of the project. ProjectWise is client/server and web portal software whose primary purpose is the organization of design data. Typically, an engineering or architectural firm would host ProjectWise on their own server and use it to collaborate and share design information with joint venture partners, clients, and subcontractors. ProjectWise offers the following features:

- Integration with Microsoft SharePoint. Project design documents can be accessed using a web browser. ProjectWise components can be added to a SharePoint portal. The SharePoint portal can also be used to provide project status information, and collaboration modules such as threaded discussions. Figure 6-13 shows the opening window of a SharePoint web portal with integrated ProjectWise data. Figure 6-14 illustrates how a 3D rendering can be displayed on the web page. Microsoft SharePoint is discussed in more detail in Chapter 7.
- ProjectWise can display Geospatial data generated by GIS software. Figure 6-15 provides a ProjectWise window showing map data.

- ProjectWise can manage and store multiple versions of a design document. It provides an audit trail for all project documents and can ensure that all users have access to the latest version of a document.
- ProjectWise can manage CAD drawings in both MicroStation and AutoCAD formats as well as Microsoft Office and Adobe Acrobat files.

ProjectWise, integrated with Microsoft Sharepoint, provides a web portal with an emphasis on the management and exchange of design documents. It illustrates how web portals can be tailored to the individual needs of the different participants in the project life cycle.

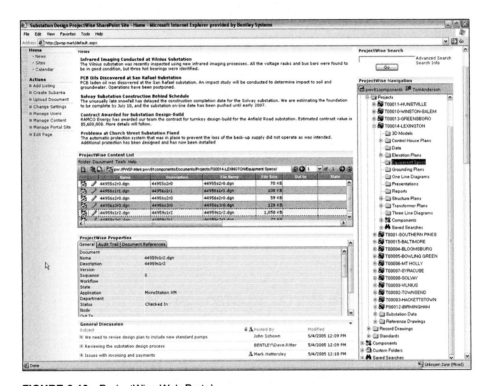

FIGURE 6-13   ProjectWise Web Portal
Courtesy of: Bentley Systems and Minnesota DOT

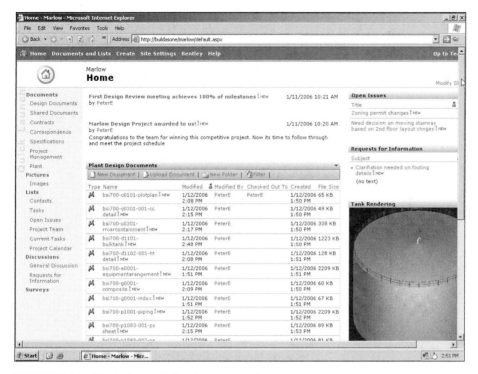

FIGURE 6-14 Displaying a 3D Rendering in the ProjectWise Web Portal
Courtesy of: Bentley Systems and Minnesota DOT

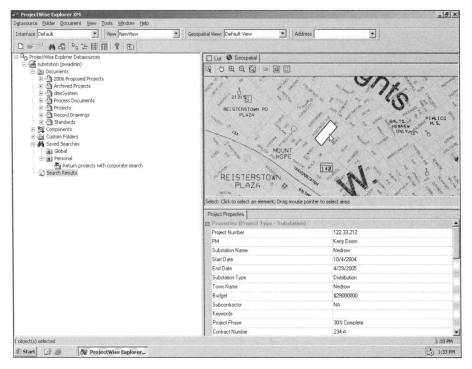

FIGURE 6-15 Displaying Geospatial Data Using ProjectWise
Courtesy of: Bentley Systems and Minnesota DOT

■ Cautions in the use of Web Portals

Web Portals are a powerful tool for managing construction projects and integrating design and construction. However, there are several management issues that must be considered when employing web portals. First, web portals are often adapted in different ways. A construction company may decide to employ a particular web portal package on all of its projects. However, often an owner requires a contractor to implement a particular type of web portal in the contract documents. It has therefore been found that construction companies must typically use different web portal software simultaneously as they conduct their portfolio of projects. Therefore, a construction firm must consider that it has personnel trained in the various types of web portals and that it provides a method of knowledge management to train employees about the idiosyncrasies of each web portal system it will be using. A construction company must also consider the establishment of standard procedures for the best ways to use web portal systems (Yeomans et al. 2004).

■ Links to Web Portal Service Providers

The following table provides a listing of the web sites for some web portal service providers. There are many different construction web portal services in the marketplace and this list is not exhaustive.

TABLE 6-1 Links to Web Portal Service Providers

Web Portal Service	Link for Information
Primavera	www.primavera.com
Constructware	www.autodesk.com/constructware
Autodesk Buzzsaw	www.autodesk.com/buzzsaw
Viecon by Bentley	www.viecon.com
ProjectWise	www.bentley.com/en-US/Products/ProjectWise/
Hard Dollar	www.harddollar.com
Meridian System Prolog	www.mps.com
Citadon	www.citadon.com
A site	www.asite.com
4projects	www.4projects.com
e-Builder	www.e-builder.net
BuildOnline	www.buildonline.com
Pro execute	www.proexecute.com
Welcom OpenPlan	www.welcom.com

■ References

Emerging Solutions, Inc. 2005. Constructware Products-Overview. http://www.constructware.com/Company/Products/default.asp (Accessed August 29, 2005).

Meridian Systems. 2005a. Prolog Manager Overview. http://www.mps.com/products/prolog/PM/index.asp (Accessed August 29, 2005).

Meridian Systems. 2005b. Prolog Pocket Overview. http://www.mps.com/products/prolog/PP/index.asp (Accessed August 30, 2005).

Yeomans, S., Bouchlagem, D., and El-Hamalawi A. 2005. Collaborative Extranet Working Via Multiple Construction Project Extranets, In *Innovations in Architecture, Engineering and Construction, Proceeding of the 3rd International Conference in Architecture, Engineering and Construction, Vol. 1.* edited by Sevil Sariyildiz and Bige Tuncer, eds, 211–222. Rotterdam: Delft University of Technology.

CHAPTER 7

Content Management Systems for Construction Management

INTRODUCTION

Content Management Systems (CMSs) are another web-based method for collaboration. For a sophisticated construction company they can be a way of developing a customized web portal. For smaller companies they can be an easy way of developing knowledge management web sites that have more capabilities than a weblog or a wiki. This chapter explores the capabilities and applications of CMS.

■ Content Management Systems

The primary purpose of a CMS is to separate web site content from site design. With the evolution of content management system software, this means that sophisticated, web sites can be constructed by users having relatively little computer knowledge, and it does not require IT personnel to enter new material or reformat the web site (Sullivan 2001). A content management system allows dynamic web sites to be constructed where new information can be easily added.

Introduction to Content Management Systems

At their most basic level, CMSs are defined as computer software that allow a user to manipulate the content on a web site. In this book we have considered several IT applications that fall under the umbrella of the basic content management system definition. In Chapter 5, weblogs and wiki were considered and we discussed their basic content management features. Importantly, construction web portals that have been discussed in Chapter 6, including Constructware and Expedition, can be considered CMS systems that have been tailored for the management of information on construction projects. Although web portals for construction are specialized CMS systems for project information exchange, it is useful to explore how construction companies can employ flexible, general-purpose CMS systems for other information and knowledge-exchange applications.

CMS can be employed in many useful ways in the construction industry. The software that is available for CMS varies broadly and can be applied for both knowledge and information management. The following is a list of the various CMS applications:

- Capture a construction firm's knowledge for future reuse. Develop and share lessons learned.
- Provide an educational tool that allows less experienced managers to access the firm's knowledge.
- Provide a web portal that organizes a firm's knowledge and information for use by company personnel.
- A web-based CMS has the potential to provide managers in the field with information about the best construction techniques.
- A CMS can be used both in the field and for home office functions like estimating and bidding. Several topical CMS could be employed within the same firm.
- CMS in their various forms can be used as a repository of knowledge and information of all of a construction company's operations, best practices, and experiences.

In this chapter we will discuss how more complex CMS systems can be used to organize information and knowledge for a construction company, and promote collaboration between members of the firm. In this chapter we will also consider the broad range of CMS software that is available. Many useful software tools and various web services are available to build content management systems without any additional investment in computer equipment. This convergence of software and web technology now makes it possible for even small construction enterprises to develop useful content management systems. Additionally, the development of these tools allows large construction organizations to consider building customized CMSs tailored to their information and knowledge-exchange requirements.

A more detailed definition of a CMS is that it is a software system that consists of two important elements. The first major element allows authors to create content for the CMS web site. A typical CMS allows the content to be created by the author without knowledge of HTML or the need for a webmaster to put the information on the web site. This allows "dynamic" web sites to be created where the content is always up to date. The second major element of a CMS is the capability of the software to compile the information input by authors and to update the web site automatically.

Basic Content Management System Capabilities

All content management systems have several basic features. Content management allows word-processing documents and scanned paper files to be automatically formatted into HTML or Portable Document Format (PDF) files for use on the web page. Most CMSs also provide some type of revision control. The revision-control feature allows content to be updated to a newer version or restored to an older version. Reversion control also tracks changes made to files by individuals. Also, a CMS system automatically indexes all data within the system and displays some form of table of contents of what documents are contained in the system. CMS systems also typically allow text searches for document keywords.

Document Management within a Content Management System

Document management is closely related to content management. Most Content Management Systems provide the capability to manage documents. Document management can be defined as the means of managing documents and images from creation to storage and dissemination to end-users. Document management includes indexing and retrieval of documents by an organized method. Two different types of document management systems can be distinguished. A document imaging system takes paper documents, scans them, and stores the image in a database. In an electronic document system it is possible to move closer to the idea of a paperless office. In an electronic document system all documents are created electronically (a letter created in Microsoft Word for example) and printed copies need not be produced.

The Content Management Systems we discuss in this chapter include document management features of the second type. In the example CMS discussed in this chapter, documents are stored in the computer system and retrieved as necessary. Document Management Systems can also include methods for providing electronic signatures on documents.

Many construction contractors already maintain simple document management systems for organizing their documents on a computer. Software applications like PaperPort (http://www.nuance.com/paperport/) allow users to organize and view common computer file formats and scanned documents. PaperPort allows scanned documents, photographs, PDF files, and Microsoft files to be viewed and organized as small thumbnails without opening the original application. Of course, a web portal or CMS typically offers many document management features. However, for a small contractor or a small project this may provide an adequate solution for organizing documents.

Possible Applications of Content Management Systems in Construction

The web portals we have discussed in Chapter 6 are construction content management systems. These systems have predefined templates for certain standard construction documents such as change orders and requests for information. The web portals also serve as a repository of documents related to particular projects.

Other CMS applications are possible beyond the limitations and topics included in a construction web portal. Perhaps most promising is the capability to create customized CMSs that store and disseminate the construction firm's knowledge. A construction web portal is primarily for the exchange of information and documents. In parallel, CMS systems that contain lessons learned, best practices and compilations of the best ways to perform construction operations and procedures can be developed. Additionally, it is possible to create a web portal that contains a rich mixture of both information and knowledge useful to managing the construction firm and individual projects.

General Benefits of Using CMS Systems

General-purpose CMS systems provide several capabilities that are useful in the construction industry. The primary benefit is that web content can be rapidly modified and updated. Users can access the latest information via the web site. This can be particularly useful on construction projects in which conditions change rapidly. Another useful benefit is that authors of material for the web page are able to access and change the web page without the intervention of a webmaster. This may be useful for smaller construction companies that do not have extensive IT capabilities. In the construction industry, some important project participants may have few computer skills. The capability to produce input for the web page without knowing any HTML or programming allows for a much broader audience of users. Finally, a CMS allows templates to be set up so standardized documents can be developed.

Ways of Implementing a CMS System

There are several ways of approaching the implementation of a content management system. The approaches vary widely in both complexity and cost. The possibilities include:

- Purchasing commercial software and installing it on your company's server. Many different software programs are available that can be installed on a server. They vary from very inexpensive programs to programs costing many thousands of dollars that are capable of handling thousands of users. This high-end software requires extensive investment in computer server equipment and a dedicated IT staff to run and maintain it.
- Using a hosted commercial application. Many commercial content management applications are available as a web service that can be subscribed to for a yearly or monthly fee.
- Installing commercial software on a web hosting service. Confusingly, it is also possible to set up an account with a web hosting service, and then you can install software on your host's server.
- Using low-cost, open-source software on your own server. Many different varieties are available. They are most suited for applications for small- and medium-sized companies.
- Installing open-source software on a web hosting service. Many web hosting services now make provisions for some open-source CMS programs to be easily installed on their own servers. It makes it possible for simple CMS systems to be set up without requiring a company to invest in computer servers.

In implementing a CMS for your company you must consider the size of your firm, and how your firm will use the CMS. You must identify if the CMS will store knowledge, information, or some combination of both. You must also consider if you want to use the CMS as a way of improving collaboration between personnel with the addition of features such as threaded discussions. In particular you must consider your firm's IT resources and capabilities. Using a CMS will require more intervention on the user's part than a simple weblog or a construction web portal that is maintained by a service provider. Resources must be made available within the firm to manage the system, control access, and to monitor the information posted by users.

Considering Possible Construction CMS Structures

Figures 7-1 and 7-2 show possible ways that a content management system can be structured to perform information and knowledge-management functions. The strength of a content management system is the ability to tailor its content to a firm's specific needs. Therefore both of these suggested structures could be modified as necessary. Perhaps a firm may consider developing a CMS that is a repository of both information and knowledge.

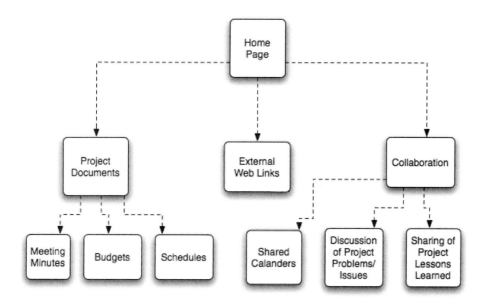

FIGURE 7-1   Content Management System Focusing on Information Management

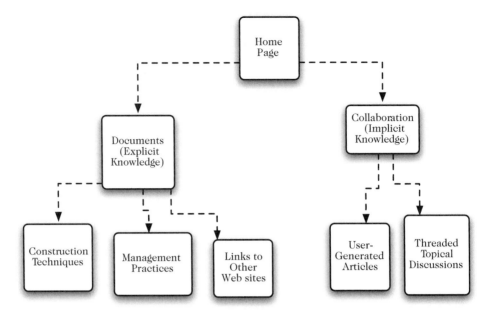

FIGURE 7-2 Content Management System Focusing on Knowledge Management

Figure 7-1 illustrates a system focusing on information management. The CMS can be set up to provide a repository on which project participants share documents. The documentation can be categorized into different areas such as meeting minutes or budgets. Collaborative features of the CMS allow for threaded discussions and information exchange between people working on the project where project issues can be recorded. Some CMSs can allow for separate workspaces to be created for individual projects, and the structure shown in Figure 7-1 could be duplicated for each project the firm conducts tied to a central home page.

Figure 7-2 illustrates a suggested structure for a simple CMS focusing on knowledge management. The CMS can be used as a repository for the firm's documentation about how to best perform different types of construction procedures, how to control construction quality, best management procedures, safety, and any other important topic. The CMS can make this documentation available throughout the firm for personnel seeking information about best practices. A CMS can also provide a way of capturing knowledge that is not documented, by providing collaboration facilities like discussion forums that allow CMS members to interact and discuss issues. Discussion and collaboration within the CMS can provide a method of informally archiving the firm's knowledge.

■ A Simple Content Management System

Some simple content management software is available that could be useful to small construction companies. CityDesk is a Windows program that generates a web site that can be transferred to a web server automatically using FTP or file copy. It operates differently than many of the CMSs that we will discuss because no software needs to be installed on a server.

Instead, CityDesk generates a web site in HTML code that can be uploaded to wherever a company hosts its web page. CityDesk automatically generates a table of contents that is the opening page of a CityDesk web site. Content can be added to CityDesk without knowledge of HTML. CityDesk has a built-in word processor that can be used to create new articles, or existing content from word-processing files can be pasted in. Pictures can also be inserted into a CityDesk article. All of the articles a user inputs are linked to the table of contents. All CityDesk articles use a standard template, which can be modified for particular applications. Using a standard template means that all content that is added to the system shares a common web design. The program also has the capability to incorporate existing HTML, PDF, and Microsoft Word documents without modification. Links can be inserted in the table of contents to this material. When new content is added to a CityDesk project, the program automatically generates a new table of contents. Additionally, the software has the capability to search all of the contained articles by keyword, so information in the system can be rapidly retrieved. CityDesk is relatively inexpensive CMS software. A starter edition that can use up to 50 files is free. A professional version costs $299 per user.

CityDesk Application for Maintenance Information and Knowledge

A prototype CMS has been developed of highway-maintenance information using the CityDesk Software (www.fogcreek.com/CityDesk/). The purpose of the system is to provide instructional documentation for maintenance crews in the field. The web site is a collection of Maintenance Operations Bulletins from the New Jersey Department of Transportation. A pavement maintenance manual from the California Department of Transportation is also included.

The opening screen of the CityDesk program is a hierarchical view of the files contained in the web site. Figure 7-3 shows the hierarchical view of the documents contained in the maintenance prototype. The site also contains CityDesk articles and files in HTML and PDF format that are linked to article files. In addition, graphics files are included in the site. Files are added to the site simply by dragging and dropping the file into the archival window.

The CityDesk software automatically creates a web index page that is the opening page of the site. A user can accept a basic template or make modifications. Figure 7-4 shows the index page for the prototype maintenance site. It was possible to add this graphic without using any HTML code. The various articles in the web site are listed on the page with links provided to the articles. If a user selects an individual article from the index, the article is displayed. Figure 7-5 shows the text of an article about maintaining curbs and sidewalks. Figure 7-6 shows the article on roadside mowing with an informative graphic file embedded within the article.

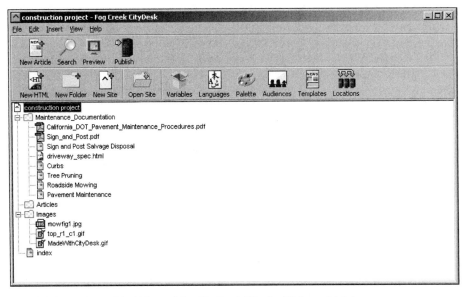

FIGURE 7-3 Hierarchical View of the CityDesk Site for Highway Maintenance

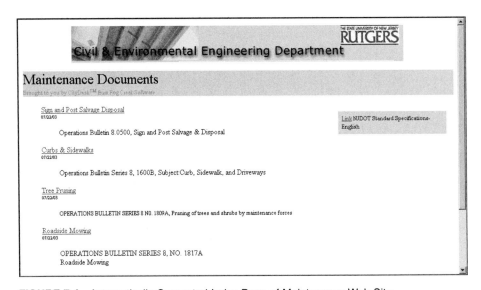

FIGURE 7-4 Automatically Generated Index Page of Maintenance Web Site

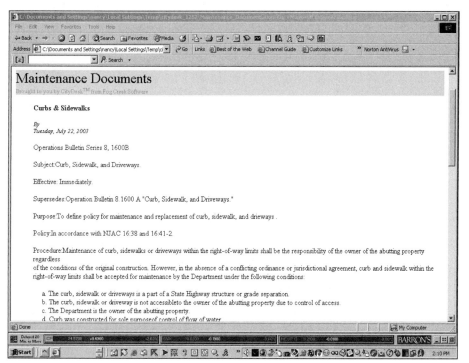

FIGURE 7-5 Text of an Article About Highway Maintenance

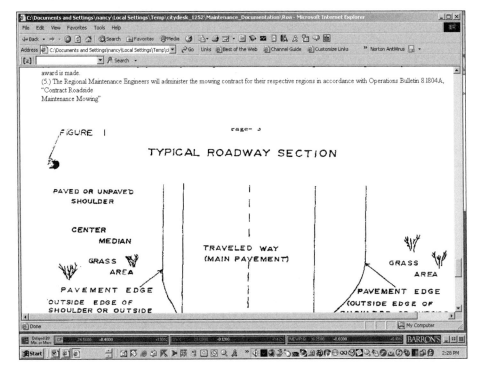

FIGURE 7-6   An Article with an Embedded Graphic

■ An Example Content Management System using Open-Source Software

Open-source software is software on which the source code can be modified or customized by the user. Open-source software is often distributed for free or for modest charges. An open-source software license may place limitations on how a user may distribute software that they have modified. A drawback of open-source software versus commercial software is that there may be little or no support available if problems are encountered.

There are many open-source CMS systems available. Widely used open-source CMS systems include PHP-Nuke, eZ Publish, and Bricolage. Many businesses find this open-source software to provide a viable solution to commercial software (Brooks 2003). In particular, open-source software is a viable alternative for small companies.

The software must be installed on a server, and typically requires software such as PHP, a computer-programming language; and MySql, a database program, to be installed on the server. However these programs are also open source and available free of charge. A good hosting service will already have this software installed on their servers. It is not necessary for you to understand how to use these programs, you only need to understand that they are required to get the CMS software to run on a server. A construction company with little computer expertise would require assistance from someone capable of setting up a server and installing the CMS and other required software. However, after the system is installed and set up on the server, construction users with less computer expertise can easily use it. A good source for information about the capabilities and computer requirements of many different open-source CMS software is provided by the web page "CMS Matrix" at cmsmatrix.org. Table 7-1 shows a listing of some popular open-source CMS software.

In this section, an example content management system will be demonstrated using the PHP-Nuke software. This section illustrates how a content management system could be set up to store knowledge and foster collaboration within a construction company.

TABLE 7-1 Open-Source CMS Software

Software Name	Web site
PHP-Nuke	www.phpnuke.org
ez Publish	www.ez.no
Bricolage	www.bricolage.cc/
Mambo	www.mamboserver.com
Typo3	www.typo3.org
Drupal	www.drupal.org

Introduction to the PHP-Nuke Content Management System

PHP-Nuke is an example of an open-source, web-based content management system that can be implemented at a relatively low cost. It requires more expertise to be used and administered than a weblog but it offers more features. It has the potential to provide a dynamic web site for a construction company on which knowledge and information can be cataloged and stored. Additionally, PHP-Nuke provides a rich set of tools for member collaboration. Small- and medium-sized firms can employ PHP-Nuke because it can be implemented on a web hosting service at relatively low costs. The capabilities of PHP-Nuke, the various ways it can be implemented, and an example application will be discussed next.

Requirements for Using PHP-Nuke

PHP-Nuke can be downloaded for free and set up to run on a server, which requires additional open-source software. A PHP-Nuke site can be hosted on a web server running the Linux software, Apache Web Server, the PHP scripting language, and the MySql database. These programs are also open-source and are available for free. These additional programs require computer expertise to install, and it is recommended that a construction company have internal IT staff or another IT professional set up the server.

Web Hosting Services for PHP-NUKE

PHP-Nuke has become a popular program for developing content management systems due to its low cost and many features. Because of this popularity, many web site hosting services are offering PHP-Nuke pre-installed so that users with little experience installing software on a server can immediately begin to use PHP-Nuke. There are many different web hosting services and they typically charge a monthly or yearly fee. The fee varies depending on the hosting service, and the amount of storage required. Web hosting services typically offer a selection of service plans. The fees range from $5 to $25 per month.

The low cost of web hosting services and the fact that the construction user is free to concentrate on developing the content management system make the use of PHP-Nuke viable for construction knowledge and information management. It is possible for a construction manager to set up and administer a complex PHP-Nuke web site without knowledge of any web programming.

PHP-Nuke Capabilities

PHP-Nuke has many capabilities (Jones 2005). It is modular in design and various features can be selected as needed. The basic content of a PHP-Nuke site is the article. Any site user can write an article. Figure 7-7 shows the form for generating an article. Like a weblog, no knowledge of HTML is required to input information in the article. The default settings of PHP-Nuke display articles chronologically in the site's home page, although this can be modified if desired. Older articles are archived and remain accessible. The site administrator must approve all articles that are input by site users before they are displayed on the web site. This allows the site content to be monitored and controlled. The use of articles can foster collaboration because other users can comment on articles. All articles are assigned to a topic that is defined by the site administrator. In a construction context, articles can be used to capture knowledge about the firm's activities. The articles can be stored by topic, making them easy to retrieve.

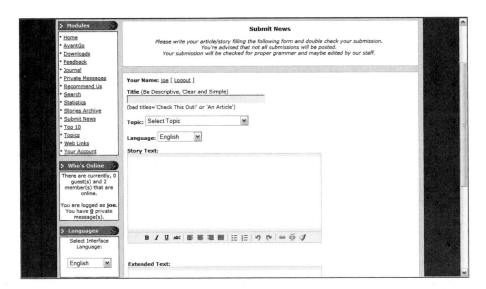

FIGURE 7-7   Form for Generating a PHP-Nuke Article
Courtesy of: PHP-NUKE-Copyright by Francisco Burzi (http://phpnuke.org)

An important feature of PHP-Nuke is the capability to provide files for downloading by site users. Using this feature of the site, documents can be organized by category and downloaded by site users. This capability allows documents containing project information and explicit knowledge maintained by the firm to be accessible to all users.

Importantly for a construction company, PHP-Nuke provides capabilities to restrict access to the site, and to content within the site. There are three main types of users that can be defined for PHP-Nuke. These are anonymous users, registered users, and administrators. For a construction company web site, typically all users would be registered. Registered users have a username and password. Administrative users have the right to add, change, and delete content from the site.

Several other PHP-Nuke modules offer useful features for managing knowledge and information. These include the ability to provide links to other web sites organized by category, forums to allow for discussions between site users, and a journal for each registered user that can be shared with other site members. Because a content management system stores so much content, and in different locations within the site, it is important to provide a search capability so articles of interest can be located. PHP-Nuke provides a search module that allows content to be searched by keyword.

PHP-Nuke provides a journal for each registered user. This journal works much like a weblog, but has some added interesting features. A user can choose to share weblog posts or his/her entire journal with other site users. This provides another method within the site where knowledge about construction procedures can be stored and shared.

A good feature of PHP-Nuke is the capability to back up the MySql database containing all of the information stored in the content management system from the server to a location of the user's choice. This would enable a construction company user to know that the CMS knowledge, information, and data would not be lost if there was a problem with the hosting company.

An Example CMS Using PHP-Nuke

To illustrate the capabilities of PHP-Nuke, a sample application has been constructed. We will discuss how the PHP-Nuke software was set up on a web hosting service. The example illustrates a CMS that might be used by a highway contractor to share and exchange information and knowledge.

Setting up a web hosting service account and installing PHP-Nuke

An account was set up on a web hosting service called Simplehost (www.simplehost.com). This service was selected because it provided the php and MySql programs on their servers (which are required to run most open-source CMS programs), and a selection of open-source content management systems that could be installed on the server. We selected the PHP-Nuke software, and automatically installed it, using installer script software provided by the hosting service. The hosting service also provided a free domain name. This is the URL that would be used to access the site. The cost for the service is $7.95 per month and includes unlimited storage space.

Articles for Discussion and Collaboration

The primary type of document that can be created by registered users of the site is articles. The default place for articles in the PHP-Nuke CMS is the center of the home page. They are listed chronologically there. Figure 7-8 shows a listing of articles written by site users. Each article is assigned to a topic area. Administrative users can create topic areas. Each topic area can have a small graphic icon associated with it. In this case, two icons are shown. One is a drawing of an asphalt-paving machine, which is attached to all articles concerned with asphalt paving techniques. A second icon showing a construction worker was associated with the safety topic. To add graphic icons requires the ability to create a GIF or jpeg file of approximately 62 x 47 pixels. After the icon is created the file must be transferred by FTP to the \images\topics directory on the web server. A system user can access a listing of articles by topic on a separate page. This allows users to review posted material by topic and article title. This page is accessed by clicking on the topics link on the left side of the home page.

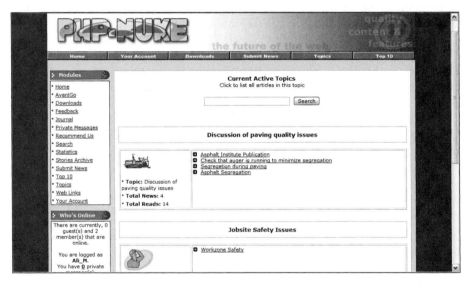

FIGURE 7-8   Listing of Articles Written by Site Users
Courtesy of: PHP-NUKE-Copyright by Francisco Burzi (http://phpnuke.org)

Weblinks to Important Sites

Figure 7-9 shows a page contained within the CMS web site that lists links to external web sites. In the example, links to DOT web sites are shown that a highway contractor would use to access applicable specifications and documents during a construction project.

Using the CMS to Share Documents

The CMS system is a good place to store documents that can be shared among construction company personnel. Figure 7-10 shows the downloads page with several categories. In the figure, categories have been created related to asphalt paving, safety, and highway-maintenance procedures. A user selects from a category of interest and is given a listing of

FIGURE 7-9   Links to External Web Sites

Courtesy of: PHP-NUKE-Copyright by Francisco Burzi (http://phpnuke.org)

downloadable documents. A listing of documents related to highway-maintenance procedures is shown in Figure 7-11. Users with administrative rights may add categories. The system is very flexible, and can allow a document repository to be created for any category.
This provides the potential for construction companies that do all types of work to tailor an
"electronic storeroom" to their specific needs. The source of the downloaded documents
may be the user's own server or other sites on the web. In setting up this example, both types
of sources were employed. Figure 7-12 shows the download form that must be filled in to
establish a downloadable document in the content management system.

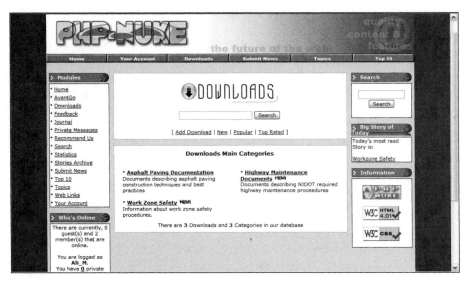

FIGURE 7-10 Download Opening Page

Courtesy of: PHP-NUKE-Copyright by Francisco Burzi (http://phpnuke.org)

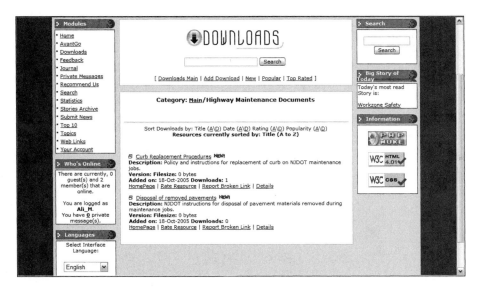

FIGURE 7-11 Listing of Downloadss Related to Highway Maintenance Procedures

Courtesy of: PHP-NUKE-Copyright by Francisco Burzi (http://phpnuke.org)

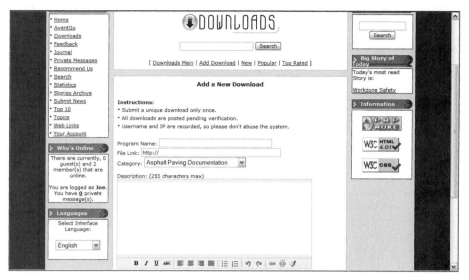

FIGURE 7-12   Download Form for the Content Management System
Courtesy of: PHP-NUKE-Copyright by Francisco Burzi (http://phpnuke.org)

■ Commercial Software For CMS Applications

There are literally hundreds of different CMS software and service providers in the marketplace. Commercial CMS software varies greatly in flexibility, features, and cost. CMS software for small enterprises can cost several hundred dollars. A powerful and flexible CMS system for a large corporation can cost several million dollars. Users also have the choice of installing the software on their own server, subscribing to a web hosting service and installing software, or using an application service provider to provide a web-based service.

Content Management Solutions from Major Software Companies

Enterprise Content Management is attracting a great deal of attention in the corporate world. The largest computer software companies provide a broad array of programs that can be tailored to the needs of very large organizations through small- to medium-sized businesses. Microsoft sells software called SharePoint Portal Server 2003 and Windows SharePoint Services. This software allows servers running Windows Server 2003 to provide users with the capability of setting up custom content management and collaboration web sites. Microsoft also provides web hosting for SharePoint, so it is possible to employ a SharePoint web site without the need to install it on a server. Used in its most basic mode SharePoint allows a CMS site to be quickly set up. However it has several powerful features that allow for customization of the CMS for more experienced users. An example of using SharePoint is given in the following section.

IBM and Oracle offer advanced tools for developing content management systems. IBM offers CMS software called Workspace Collaboration Services and Lotus Notes. The IBM offerings are broad, and a construction company should seek professional IT assistance to find the right solution for their company. Oracle's software is called Oracle Content Management TDK. It appears to be a more complicated software for building customized CMS systems and would probably be appropriate for large construction companies with IT capabilities.

Several software companies specialize in high-end content management solutions. These companies include Vignette Corp.; Interwoven; Documentum; Inc., BroadVision, Inc.; and FileNet Corp (Sullivan 2001). Software of this type is aimed at large companies wanting to provide communication and collaboration for large numbers of users widely dispersed geographically. The literature for these software companies often mentions Enterprise Content Management, indicating that the software technology is appropriate for storing company-wide knowledge and information. These content management systems typically provide powerful capabilities for searching and organizing documents. For example, Vignette can provide automated clustering and taxonomy generation and can group documents together containing similar concepts to enable searching by context, or subject (Vignette). The web sites for some of the software and services discussed in this section are shown in Table 7-2.

TABLE 7-2 Commercial Content Management Systems

Windows SharePoint Services	http://www.microsoft.com/windowsserver2003/technologies/sharepoint/default.mspx
Windows SharePoint Services hosting at bCentral	http://www.microsoft.com/smallbusiness/online/collaboration-software/sharepoint/detail.mspx
IBM Workplace Collaboration Services	http://www.lotus.com/products/product5.nsf/wdocs/workplacehome
Oracle Content Management SDK	http://www.oracle.com/technology/products/ifs/index.html
Vignette	http://www.vignette.com/
EMC Documentum	http://www.documentum.com/
BroadVision	http://www.broadvision.com/
FileNet	http://www.filenet.com/

Commercially Available Hosted Systems for Small Contractors

Many other commercial content management solutions are available. Some simple content management systems are provided as hosted web services. Good examples of this type of service are HyperOffice (www.hyperoffice.com), Basecamp (http://www.basecamphq.com/), Backpack (http://www.backpackit.com/), and ContactOffice (www.contactoffice.com). For example, HyperOffice is an easy-to-use web-based application that includes capabilities like:

- Document management. HyperOffice allows documents to be stored, tracked, shared, and allows older versions of documents to be stored.
- Online calendars. Shared group calendars can be created.
- A task manager that allows shared tasks to be created for a group. Figure 7-13 shows the web form that allows a collaborative group to be created.

- A central point for exchanging files can be established for a group. Figure 7-14 shows a sample file-sharing window that could be set up to exchange bidding documents and project details.
- Online discussion forums to promote collaboration.

This software is simple to use and is completely web based. The CMS system is predefined and does not require any sort of a web site to be constructed by the user, making it ideal for small construction companies that simply want a better way to organize their documentation and set up collaboration between personnel. The service is relatively inexpensive at about $18/month for 2 users and about $1,370/month for 250 users.

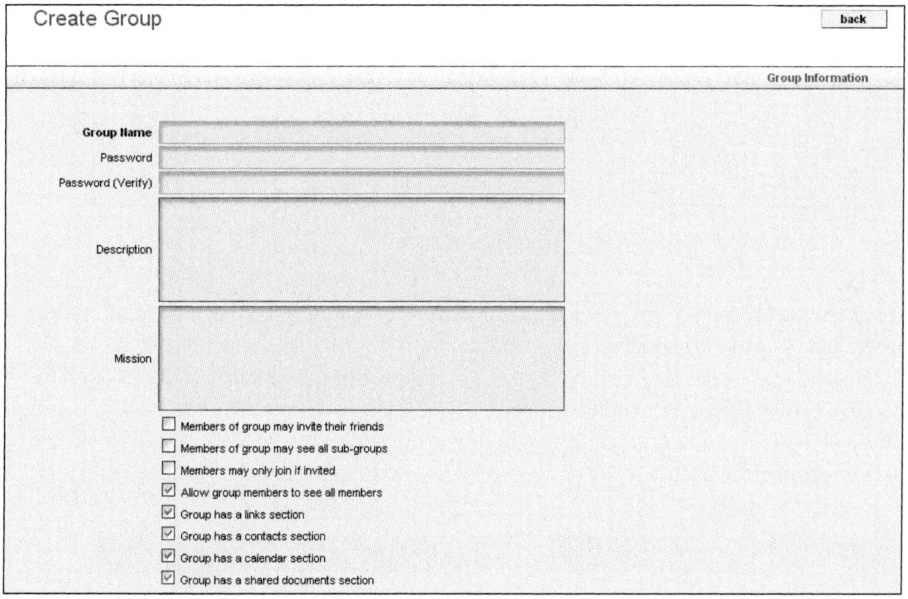

FIGURE 7-13 Creating A Group Using HyperOffice

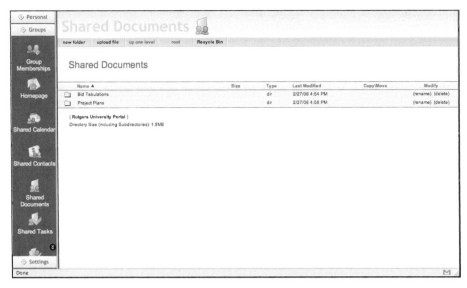

FIGURE 7-14   Web-based File Exchange Using HyperOffice

The Backpack web service provides a useful feature by giving the user an e-mail address that he or she can use to automatically post information on a Backpack web site. Figure 7-15 shows a Backpack page with two e-mails about paving problems that have been automatically posted to it. Note the page's e-mail address, which is shown below the messages. This feature can allow construction managers in the field with wireless Internet to rapidly post and archive important project events to be shared with other members of the project team.

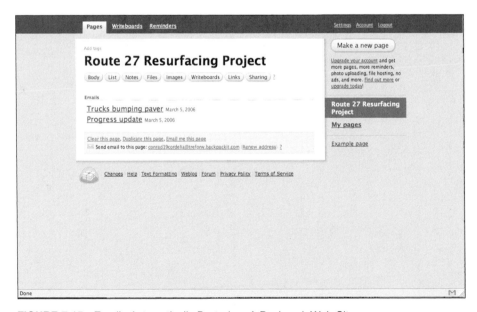

FIGURE 7-15   Emails Automatically Posted on A Backpack Web Site

An Example Using SharePoint Services

A sample CMS for a construction company was developed using SharePoint Services. A SharePoint Services account was set up through the Microsoft bCentral hosting service. An introductory account was set up that provided SharePoint Services hosted on Microsoft's servers for $39.95/month. The account allows 30 users and provides 200 MB of disk storage. There are other accounts available that add more users and storage.

Details of Using SharePoint

In this example, we show how SharePoint Services can be used to set up a simple customized web portal that focuses on the exchange of information for a school construction project. SharePoint is an interesting software system because at a basic level it allows a user to rapidly set up a content management system with little computer knowledge. However, it has many advanced features that can allow a SharePoint web portal to be highly customized.

Because this is a Microsoft product, SharePoint services is integrated with the Microsoft Office suite of programs. Microsoft Word and Excel files are easily incorporated into the content management system. Some of the basic features of SharePoint are:

- The ability to set up a web page or web site without knowledge of HTML or programming.
- A SharePoint site can contain a library of documents.
- A picture library can be created using SharePoint.
- Issue lists can be created.
- Discussion boards can be set up to provide collaboration between site users.

An advanced feature of SharePoint is the ability to set up various workspaces in the content management system to allow groups of users to collaborate. In particular it is possible to set up a meeting workspace that includes an agenda, objectives, and shared documents for invited participants. The software also allows additional sub sites to be set up containing any desired content, including the use of versioning and archival storage. Using Microsoft Infopath to create the forms, it is also possible to include and fill out forms in SharePoint Services (Londer et al. 2005).

SharePoint Example for Construction Information Exchange

An example SharePoint site focuses on information exchange for a school construction project. Figure 7-16 shows the opening page of the web site, which was set up to inform users of events and to provide folders where documents for schedules, budgets, and cost reports are stored. Figure 7-17 shows the document folder for project schedules. Notice that the Excel icon appears next to each schedule showing what type of Microsoft Office document it is. It is interesting to note that SharePoint is set up to operate by accepting Microsoft Office files that have been created on a user's personal computer. This contrasts with the example of PHP-Nuke above where users can enter content as articles by filling out a web-based form. Figure 7-18 shows the ability of SharePoint to store and display photographs. In the example, progress photographs are stored in the content management system.

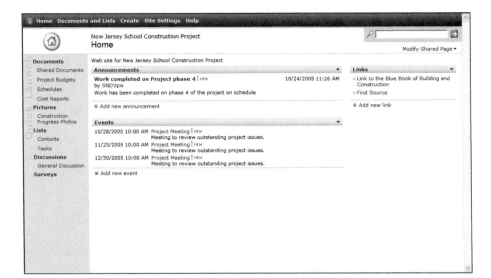

FIGURE 7-16   Opening Page of SharePoint Services Web Site

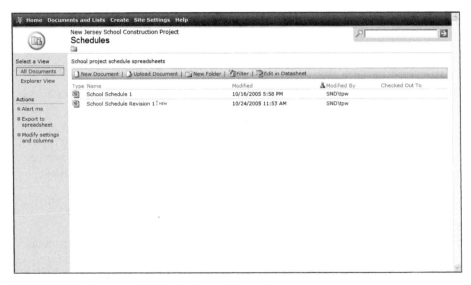

FIGURE 7-17   Document Folder for Project Schedules

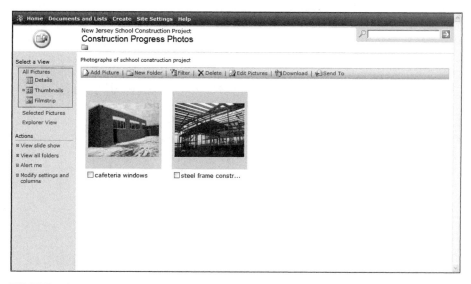

FIGURE 7-18 Displaying Photographs

The workspace for a project meeting is shown in Figure 7-19. This figure shows the meeting agenda, documents related to the meeting, and the invited participants. This workspace is automatically generated by the software as a sub site to the main project page. This example shows how the SharePoint Services software provides an easy way to set up a CMS and how it is integrated with the various types of Microsoft Office documents. For a small- or medium-sized construction firm that already uses Microsoft software extensively, SharePoint services are a good way of providing a CMS for collaboration and document exchange (Londer et al. 2005).

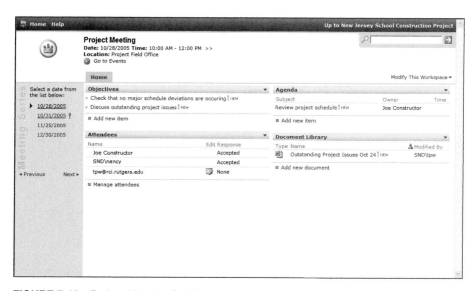

FIGURE 7-19 Project Meeting Subsite

■ Conclusions

A content management system can improve collaboration within the construction company. They vary from simple to highly complex systems. They provide more flexibility and features than the weblogs and wiki in Chapter 5. In Chapter 6 we discussed construction web portals that are appropriate for many companies. The content management systems discussed in this chapter are for companies wishing to have more customization of the content stored in their computer systems. Content management systems are also appropriate for companies interested in going beyond project information exchange, and that wish to construct ways to capture a construction company's knowledge. Due to the great interest in knowledge management shown by many companies, it is anticipated that software and technology for content management systems will continue to improve and evolve.

■ References

Brooks, J. Open Source Provides Viable CMS Option. *eWeek.com*, June 1, 2003, http://www. eweek.com/article2/0,1759,826084,00.asp (Accessed October 15, 2005).

Jones, D. *PHP-Nuke Garage*. 2005. Upper Saddle River, NJ: Pearson Education, Inc.

Londer, O., Bleeker, T., Coventry, P., and Edelen, J. 2005. *Microsoft Windows SharePoint Services Step by Step*. Redmond, WA: Microsoft Press.

Sullivan, B. 2001. Content Management for the Masses. *USC Annenberg Online Journalism Review*, June 28. http://www.ojr.org/ojr/technology/1015018005.php (Accessed August 25, 2005).

Vignette Corporation. Vignette Taxonomy and Advanced Search. http://www.vignette.com/contentmanagement/0,2097,1-1-1928-4149-1968-4326,00.html (Accessed October 24, 2005).

CHAPTER **8**

Online Bidding and Online Plan Rooms

INTRODUCTION

Traditionally, construction contractors needed to travel to plan rooms to view project documents when they considered bidding on a project. Fees were often charged by agencies for contractors to purchase sets of contract documents. When submitting bids, contractors often needed to travel to locations far from their office to submit bids. Bid prices were typed on proposal forms, a process prone to mistakes. Now web-based services exist that allow contract documents and plan sheets to be downloaded over the Internet. The bidding process can also employ the web by allowing contractors to submit their bids online, eliminating the need to travel to another office to submit a bid, and eliminating many of the errors that are possible when submitting a traditional proposal form. This chapter discusses both online plan rooms and online bidding. An example of this is Bid Express, a sophisticated system used by many state departments of transportation (DOTs) to distribute project plans and receive bids electronically.

■ Online Plan Rooms

The purpose of plan rooms has always been for owners, designers, and general contractors to make plans available to potential project bidders and suppliers. Now, online plan rooms are web-based services for which a user can access project plans and information without leaving his or her office. Users of an online plan room, including subcontractors and suppliers can view, download, and print selected project files. Benefits of using project plan rooms include increased bidding efficiency on new projects, easy access to information throughout the project lifecycle, a new source for new project leads, and the ability to be notified of new project availability by e-mail (Jurewicz, 2002). Some plan room services are run by local reprographers and include the ability to purchase hard copies of plan sets. Some construction contractors are not yet fully comfortable with the paperless construction office and prefer to receive hard copies.

There are several national plan room services. These services allow an owner or designer to create a virtual plan room that allows interested contractors to view the project before bidding. These services include:

- BB-Bid (*www.thebluebook.com*)
- Dodge Plans (dodge.construction plans.com)
- iSqFt (*www.iSqFt.com*)

The iSqFt service provides electronic plan rooms associated with the Associated General Contractors in many markets. Typically, electronic plan rooms offer their services for a fee or a mixture of free and fee-based services.

BB-Bid is owned by Contractors Register, Inc. that publishes the *Blue Book of Building and Construction*. Construction project participants can use BB-Bib in several different ways:

- as a marketplace where owners can advertise new contracts to bid, and general contractors can seek quotations from subcontractors
- as a place where subcontractors and material and equipment vendors can seek leads for new business
- as a means to establish an access-controlled online plan room, which can contain both plans and specifications

BB-Bid also offers automated invitation to bid e-mail messaging that can be sent to general contractors, subcontractors, and material vendors to inform them of the availability of a project. The types of messages that can be sent using BB-Bid include invitation to bid, requests for information, request for quotation, and notice of addendum. This capability is linked to the Blue Book and the user can select to have his or her project advertised to Blue Book vendors (Contractors Register, Inc.). Clearly, the establishment of electronic plan rooms is a convenience for everyone involved in the construction process because it reduces printing and travel costs and allows projects to be advertised to a larger group of potential bidders. Online plan rooms ease the transmission of new project information to interested parties. The next step is to consider using the web to receive online bids from contractors.

■ Online Bidding

Online bidding allows bids to be submitted electronically using the Internet. Many owner organizations are now requiring the use of online bidding for their projects and its popularity continues to grow. Some owner organizations are now accepting only online bids and will no longer accept paper bids. Therefore it is important for everyone in the construction industry to learn how these systems work.

Online bidding exchanges allow bidders to receive and submit plans and bids electronically. Online bidding has many benefits for contractors. These include the capability of easily obtaining project plans and documentation at a low cost, and reduced bidding costs because it is not necessary for anyone to travel to an owner's office to submit a bid.

■ Online Bidding Example

The state transportation agencies in the United States are in the process of standardizing an online bidding system for their highway and transportation infrastructure projects. Projects undertaken by the Department of Transportation (DOT) are in the public domain and therefore typically competitively bid. Traditionally, these projects are unit price contracts that require the submission of a proposal form that gives a unit price and total price for each bid line item as well as a total price for the entire project. The following is an example of the operation of this system.

Bid Express

Bid Express is an online bidding exchange for highway construction projects. It is currently used by approximately 20 state highway agencies to provide project plans and documentation to contractors and to receive bids from highway contractors. The expanding use of online bidding shows how e-commerce use is expanding into all types of business transactions and that a successful construction company must be able to use these tools to compete successfully.

Bid Express uses several security features to insure the security of the submitted bid information. Strong encryption methods are used which make files unreadable if they are intercepted. A state DOT does not receive electronic bids until the bid opening. Before the bid opening, the bid data is maintained in a "locked box" by Bid Express.

■ The Functions of Bid Express

Bid Express has three main functions. The first function is to provide plan sheets and other documentation in digital format for construction contractors preparing bids. The second function is to provide a method where contractors can obtain an electronic version of the project proposal form, and the third function is to receive paperless electronic bids from construction contractors.

Digital Plan Room

Bid Express maintains a digital plan room. All projects that are being advertised for bid by a state DOT can be selected. Each project presents a listing of PDF files of each plan sheet. The user can select all of the project plan sheets for downloading or select custom groups of plan sheets. Figure 8-1 illustrates a Bid Express web page that shows thumbnails of plan sheets for a project that are available for download. A user can select an entire project to be downloaded or use the checkboxes shown in the figure to select individual plan sheets.

One- and Two-way Bidding

Bid Express can work in two different ways. It can submit both one-way and two-way bids. With a one-way bid, the contractor downloads the project proposal from the Bid Express web site and fills it in using the Expedite program on their personal computer. Then they produce a hard copy of their bid and a copy of the bid on a CD and submit it by mail to a

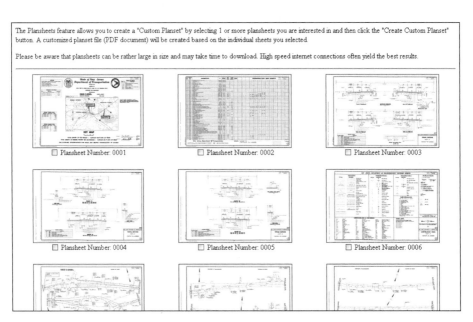

The Plansheets feature allows you to create a "Custom Planset" by selecting 1 or more plansheets you are interested in and then click the "Create Custom Planset" button. A customized planset file (PDF document) will be created based on the individual sheets you selected.

Please be aware that plansheets can be rather large in size and may take time to download. High speed internet connections often yield the best results.

☐ Plansheet Number: 0001   ☐ Plansheet Number: 0002   ☐ Plansheet Number: 0003

☐ Plansheet Number: 0004   ☐ Plansheet Number: 0005   ☐ Plansheet Number: 0006

FIGURE 8-1 Plan Sheet Thumbnails
Courtesy of: Info Tech, Inc.

DOT. Two-way bidding allows the contractor to download the project information electronically, and then submit the bid to the state DOT electronically through Bid Express by uploading the bid forms filled out in Expedite, which is provided free to users of the Bid Express service.

Using the Expedite Program with the Bid Express Service

The Expedite software is used to view the project proposal information downloaded from the Bid Express web site. It is then used to fill out the proposal form, insure DBE compliance, and obtain a bid bond for a project. When the proposal form has been filled out with the bid data, it is then used to create a secure version of the completed bid documents for submission to Bid Express. In infrastructure construction the proposal form lists each bid item and it must be filled in with unit prices, the total price for each line item and the total project bid price.

Downloading Bid Forms from the Web

Bid proposal forms are selected from the Bid Express web site and downloaded to the user's personal computer for processing using Expedite. Figure 8-2 shows how the available bid data is arranged by letting date on a state DOTs Bid Express page. Each state has its own set of Bid Express web pages. Figure 8-3 shows the listing of projects for a single bid letting date. The primary form that is accessed using Expedite is the bid proposal form. The contractor uses the proposal form to insert their bid prices. If an individual project is selected from the listings for a letting date its proposal form is displayed on Bid Express. This is illustrated in Figure 8-4. In addition, files that may be downloaded and links to additional project-specific information are available from this page. The upper right-hand corner of

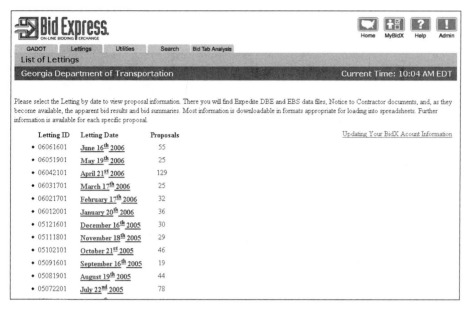

FIGURE 8-2 List of Lettings
Courtesy of: Info Tech, Inc.

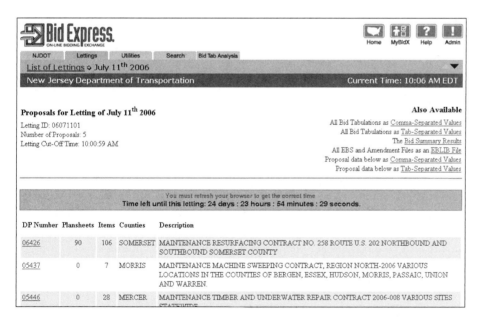

FIGURE 8-3 List of Proposals for a Letting Date
Courtesy of: Info Tech, Inc.

Figure 8-4 shows the files that are available for download including the bid proposal form with an .ebs extension that is readable in Expedite.

DP Number 06426

Date Generated: June 15th 2006
Date Revised: N/A
Items: 106
Sections: 1
Amendments: 0
Highway Number:
Counties: SOMERSET

Also Available

The Plansheets Documents
NEW! Bid Tab Analysis
REQUIRED: Expedite Data File 11jul426.ebs
Special Provisions File
Notice to Contractors
REQUIRED: Bid Bond Form
Item data below as Comma-Separated Values
Item data below as Tab-Separated Values

MAINTENANCE RESURFACING CONTRACT NO. 258 ROUTE U.S. 202 NORTHBOUND AND SOUTHBOUND SOMERSET COUNTY

Create a Quotation Form

Section 0001: ROADWAY

Line	Item	Alt	Description	Units	Price	Quantity
1	20011C21E		PERFORMANCE BOND AND PAYMENT BOND	LS		1.00000
2	20011E21C		CONSTRUCTION LAYOUT	LS		1.00000
3	20011E31E		FIELD OFFICE TYPE B SET-UP	U		1.00000
4	20011E31F		FIELD OFFICE TYPE B MAINTENANCE	MO		10.00000
5	20011E31G		TELEPHONE SERVICE	LS	3600.00000	1.00000
6	20011F21C		MATERIALS FIELD LABORATORY SET-UP	U		1.00000
7	20011F31C		MATERIALS FIELD LABORATORY MAINTENANCE	MO		8.00000
8	20011F41C		CURING AND STORAGE FACILITY SET-UP	U		1.00000
9	20011F51C		CURING AND STORAGE FACILITY MAINTENANCE	MO		8.00000

FIGURE 8-4 Proposal Form in Bid Express
Courtesy of: Info Tech, Inc.

Project proposal forms can be viewed online through the Bid Express web site, but prices can only be entered in Expedite, which provides a higher level of security because it is possible that computer hackers could view bid prices if they were input directly via a web page.

Entering Bid Prices

Figure 8-5 shows the Expedite proposal form on a personal computer. Initially all of the prices for the proposal items are blank and the file folders on the left are red, indicating they are incomplete. When all of the required information for a project has been filled in, the folders turn to green. The information that is input in Expedite includes the bid prices, the DBE commitments, and Bid Bonds (These are purchased online and the procedure is described below.)

Each project item is listed in the proposal form along with the unit and quantity. The contractor enters prices on this form. Error checking is available within the Expedite program to insure that no items are left blank. The Expedite program automatically "runs the extensions" multiplying unit price by quantity to compute the item price automatically. The program also automatically sums item totals to produce the total bid price.

Entering DBE Commitments

Federal and state laws require contractors to subcontract a certain percentage of their project to Disadvantaged Business Enterprises (DBEs). States typically establish DBE goals for their projects and the contractor must certify in her bid submission that these DBE goals have been met for the project. The listing of certified DBE contractors is included in

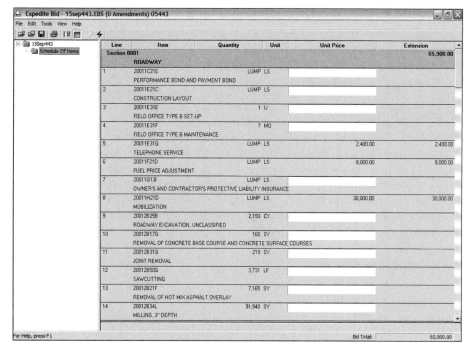

FIGURE 8-5 Proposal Form Downloaded to Expedite

the Expedite program (The Expedite program is customized for each state DOT). A contractor selects the contractor who has been hired and clicks on the bid items that this subcontractor will be responsible for. A screen in the DBE portion of Expedite indicates what amount of the DBE goal has been fulfilled.

Bid Bonds

Bid Express also automates the process of obtaining bid bonds for a construction project. Typically an owner requires a bidder to obtain a bid bond to protect the owner from the financial loss he/she may suffer if the low bidder fails to enter into a contract (Halpin & Woodhead 1998). The Bid Express system is integrated with three surety companies that maintain online web sites where contractors can obtain bid bonds. The Bid Express program is able to access the surety's database and confirm electronically that a contractor has obtained the proper bonding. This confirmation must be obtained before a bid can be uploaded to Bid Express.

Bid bonds can be obtained online from several vendors, including Surety 2000 and Insure Vision. Surety provides a bid bond verification number that is entered in the bid bond section of Expedite. Expedite then connects to the bonding company through Bid Express and verifies that a bid bond has been obtained. This electronic verification is submitted to the DOT along with the bid.

Digital Signatures

Submitting a bid using Bid Express requires a user to have an electronic signature on file. This electronic signature can be established for a one-time fee of $100. A form is generated by the Expedite program that a user fills out, signs, and mails to Bid Express. In return, the user receives the necessary passwords required to submit a bid to the Bid Express system.

Subcontractor Quotations

The Bid Express system provides a method for subcontractors to develop quotations to be sent to prime contractors. The "Create a Quotation Form" link is shown in the center of Figure 8-4. If this link is selected, the bid proposal form is displayed with a check box next to each bid item. Subcontractors are able to click on project items that they wish to undertake and can then download the information to their computers as a CSV (Comma Separated Value) file that is readable in a spreadsheet program. The subcontractor can then fill in the bid prices in the spreadsheet and e-mail or fax the quote to the prime contractor.

Modifying and Removing Bids

The Expedite software provides the capability to both modify a bid and to withdraw a previously submitted bid. Bid Express has found through experience that contractors typically upload a "safe bid" during the bidding process (Usually one or two days before the scheduled bid opening). They can then modify this initial "safe" bid as needed during the later stages of the bidding process. The "safe bid" gives the contractor the security that he/she has submitted even if some later events prevent a final fine-tuning of the bid.

■ Using Bid Express to View Bid Results

Bid Express can be used to view the results of bid openings. The "apparent bid results" can be displayed using Bid Express. These results are available instantaneously as the letting is occurring and the user needs only to refresh the web browser to view them. Figure 8-6 shows a sample of bid results. Bid summary information, a final version of the bidding results, is available after the bid is processed through a DOT's letting and awards system (This process may take up to 14 days to complete).

■ Bid Tab Analysis

An optional feature of Bid Express is "Bid Tab Analysis." This feature enables a contractor to view historical bidding information by item and the contractor that bids on the item. The search can be filtered by date or item type (Info Tech 2004). Figure 8-7 shows a display of bid tab data for "Asphalt Joint and Crack Sealer" for a selected time period. The web page lists the contracts that the item was used in and the average bid, high bid, and low bid found in each project. This feature puts bid results in an easily accessible format for contractors preparing bids and reduces the time necessary to access historical information when preparing a new bid.

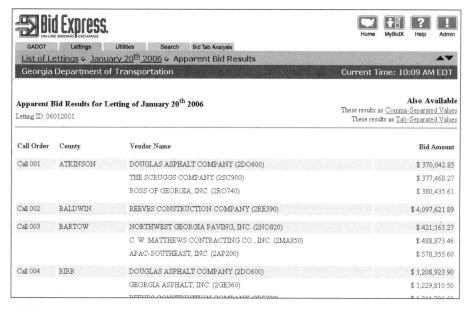

FIGURE 8-6 Bidding Results
Courtesy of: Info Tech, Inc.

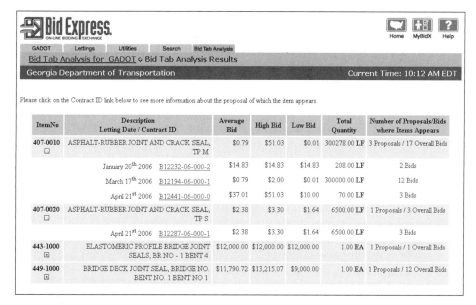

FIGURE 8-7 A Bid Tab Analysis
Courtesy of: Info Tech, Inc.

Costs of Using Bid Express

A basic subscription to Bid Express costs $35/month. This allows unlimited downloading of letting information for all of the state DOTs. It costs an additional $95/month to access the electronic plan room and have the ability to download unlimited PDF files of the plan sheets. The cost savings using this type of system can be large, considering contractors typically pay from $200- to $4,000 per month for plan sheets. Contractors using two-way bidding, submitting their bids via Bid Express, pay $15/month/state for unlimited submittals.

Reverse Online Auctions

The use of online bidding can be extended beyond the traditional sealed bid auction. In the method employed by Bid Express, traditional bidding methods have been employed in which a bidder's price is revealed to other bidders only at the time of the bid opening. Web-based technology is now allowing transparent reverse auction bidding whereby each bidder's price is displayed to other bidders, and bidders may modify their bids based on the behavior of other bidders. This method of online bidding can be applied to two areas of construction. They are bidding for actual projects, and procurement of materials and supplies. There are many online auction service providers. Two examples are HedgeHog (www.hedgehog.com) and iasta (www.iasta.com).

Reverse auction online bidding for construction projects typically works in the following manner (Canadian Construction Association 2001):

- The owner develops a list of pre-qualified contractors and invites them to participate in the reverse auction. Drawings and specifications, along with instructions on how to participate in the reverse auction, are provided to the bidder in advance of the event.
- An online, Internet-based auction is scheduled and conducted on behalf of the owner by a third-party, web-based service provider, with all bidders participating simultaneously, with a specified start and closing time. A reserve price may be stipulated by the owner, usually based on an engineer's estimate.
- The identity of the bidders is confidential during the bidding process.
- The bidders submit initial prices once the reverse auction begins. The submitted bid prices are communicated back to all participating bidders, with the bidder being told of their own ranking relative to others. The bids are ranked in ascending order and the lowest priced bid is the highest ranked.
- Bidders can re-submit lower prices as often as they wish up to the specified closing time, with new rankings communicated back to all bidders as new bids are made.
- The auction closes when no more new bids are placed and the closing time expires.
- All bidders are immediately notified of the final bid rankings, with only the dollar value of the winning bid disclosed.
- The owner is notified of the bidding results by the auction service provider.
- The owner contacts the winning bidder (the lowest bid) to complete the formal award of the contract.

The use of reverse online auctions for project bidding is controversial. Many contractor organizations argue that the reverse auction process encourages contractors to initially submit inflated prices and that the final low bid price obtained from the auction is not guaranteed to be the best project price. Alternatively, it may be possible that a highly competitive market causes contractors to submit unrealistically low prices that will require them to construct the project at low quality.

The U.S. Army Corps of Engineers has conducted a pilot program in the use of reverse auctions and has concluded that their primary usefulness is for commodities procurement. The Corps concluded that because construction projects are unique, they are best procured through the use of sealed bids (www.enr.com 2004).

Reverse auctions on the Internet impact contractors in two ways. First, a contractor must be aware of the issues discussed above when considering bidding on a project for which an online reverse auction will be used. Second, contractors may be willing to consider using reverse auctions for the procurement of materials and supplies for a project. Online reverse auctions are already being used in many industries for these purposes. Perhaps, reverse online auctions also show the potential for new technologies to have a disruptive influence on long-established practices in the construction industry.

■ The Future of Online Bidding and Online Plan Rooms

This demonstration of the use of the Bid Express system indicates that online bidding can provide benefits for construction contractors. Benefits to contractors include the ability to easily submit bids to a geographically distant owner, the error checking features of the software when filling in the proposal form, and the ability to easily obtain addenda and plan sheets. There will be increasing expansion of online bidding for all types of construction projects.

The discussion of reverse auctions shows that advancements in IT are allowing new forms of bidding procedures to be used. A contractor must carefully weigh the positives and negatives of these new methods of online bidding before participating. New information technology not only can produce greater efficiencies, but as the reverse auction discussion points out, can disrupt long-established bidding methods.

■ References

Canadian Contractors Association. 2001. *A Contractor's Guide to Reverse Auctions*. http://www. cca-acc.com/news/committee/rag/rag-contractor.pdf (Accessed March 1, 2006).

Contractors Register, Inc. *BB-Bid Frequently Asked Questions*. http://www.thebluebook.com/faq2. html (Accessed September 23, 2005).

ENR.Construction.Com. 2004. *Corps Says Reverse Auctions Not for Construction Services*. *McGraw-Hill Construction*, August 2, http://www.construction.com/NewsCenter/Headlines/ ENR/20040802f.asp (Accessed February 15, 2006).

Halpin, D. and Woodhead, R. 1998. *Construction Management*. New York: John Wiley & Sons.

Info Tech. 2004. Bid Express User's Guide. *http://www.bidx.com/main/info/faq/usersguide.pdf* (Accessed September 14, 2005).

Jurewicz, J. 2002. *The Emergence of Online Plan Rooms. Electrical Construction & Maintenance*, April 1. http://www.ecmweb.com/mag/electric_emergence_online_plan/ (Accessed September 23, 2005).

CHAPTER **9**

3D, 4D, and 5D CAD Applications in Construction

INTRODUCTION

New computer technology is emerging that has the potential to revolutionize how construction is planned

and executed. It is now possible to produce complex three-dimensional (3D) visualizations of construction

projects that are linked to schedule and estimate information. Recently, there has been much interest in

the use of four-dimensional (4D) and five-dimensional (5D) CAD models for use in designing, planning,

and scheduling construction projects. In this chapter we will explore the uses of 3D visualizations in the

construction industry to aid engineers, architects, and construction contractors.

What are 3D, 4D, and 5D CAD Programs?

Computer-Aided Design (CAD) programs originally were used to generate 2D plans as these programs. A 3D CAD program allows a design to be represented as a 3D image. A 4D CAD program is a 3D image of a design that is simulated over time. In a construction context, a 3D drawing of a building is linked to a schedule for the construction of the building. Users of a 4D CAD program are able to view a simulation of how a construction project evolves over time. Five-dimensional CAD programs have recently emerged. These programs couple a 3D visualization of a project, the project schedule, and project cost estimates. In Chapter 4 we discuss software for project scheduling. What is now emerging are programs that integrate a 3D CAD model of a building with CPM and line-of-balance scheduling techniques.

■ 3D CAD Software

There are many different CAD packages for architectural and civil design. Popular CAD programs such as AutoCAD and MicroStation were originally used to produce 2D project plans. These programs are widely used by designers and constructors to produce project plan sheets on the computer. Two-dimensional plans have several limitations that have prompted the use of 3D CAD in construction. These limitations include:

- 2D plans can be difficult to visualize.
- The 2D plans may require extensive interpretation by a user.

The CAD programs have evolved considerably over time and now offer powerful 3D visualization features.

These applications can provide 3D plans and renderings. Three-dimensional plans can be used from the initiation of the design through construction. Three-dimensional CAD is employed by designers including architects and engineers to visualize designs, and to identify conflicts. Three-dimensional drawings can also be used during construction by construction contractors to interpret complex designs. The benefits of applying 3D drawings to the construction process include (Cory 2001):

- Checking clearances and access
- Visualizing project details from different viewpoints
- Using the model as a reference during project meetings
- Constructability reviews
- Reducing interference problems
- Reducing rework

Identification of conflicts has been most extensively applied to complex process plants where it may be difficult to visualize how networks of pipes may conflict with 2D drawings. The use of 3D plans has naturally focused on process plants and complex commercial buildings. However, the use of 3D plans is now being extended to infrastructure design such as highways and railroad alignments.

The Bentley GEOPAK Civil Engineering suite can be used for 3D modeling of highway designs. It can be used to detect conflicts between separate elements of a design that have been generated in 2D. Figure 9-1 shows the 2D design of a retention pond, and Figure 9-2 shows the 2D design for an adjacent roadway. Examination of the two plan sheets does not immediately identify an error in grading elevations that causes a conflict. However, a 3D drawing, shown in Figure 9-3, combines the data from the two 2D drawings and shows that there is a conflict and a redesign is needed. The use of 3D CAD enables a design error to be caught during the design phase rather than requiring a more costly change order during construction.

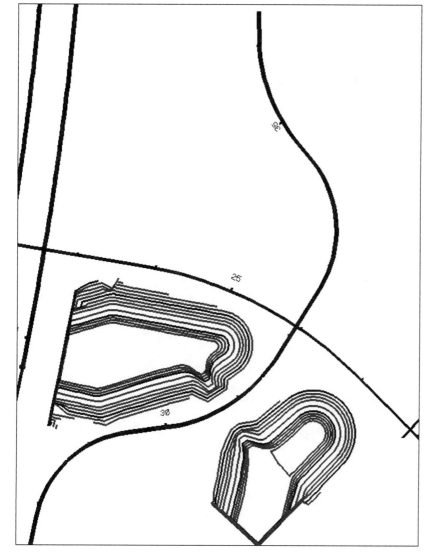

FIGURE 9-1 2D Design for a Retention Pond
Courtesy of: Bentley Systems and Minnesota DOT

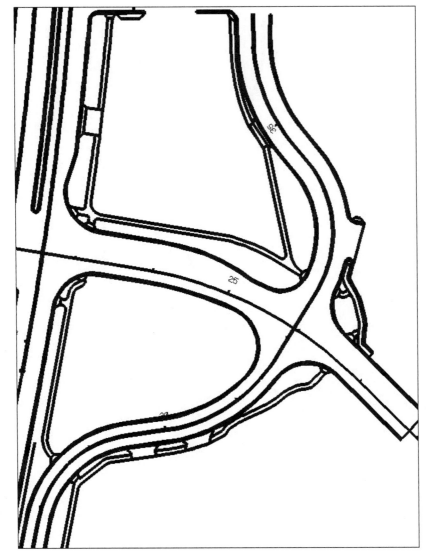

FIGURE 9-2 2D Design for Roadway Adjacent to Pond
Courtesy of: Bentley Systems and Minnesota DOT

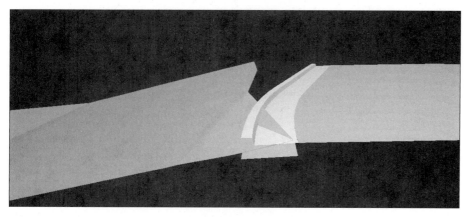

FIGURE 9-3 3D Visualization of Design Conflicts
Courtesy of: Bentley Systems and Minnesota DOT

Some Popular 3D CAD Software

The software that is available for computer-aided design is continuously evolving and is becoming increasingly sophisticated. We will review some of the best-known CAD software and how it can be employed to generate 3D CAD models. Three-dimensional CAD differs from traditional plan sheets by providing a virtual prototype of a completed building or subassembly of a building. There are three types of 3D models that can be generated by CAD software. Wire frame modeling is the simplest and shows the shape of design elements using interconnected line elements that describe edges and corners. The next level of 3D CAD is surface modeling where the outside geometry of an object is defined. The most sophisticated form of modeling, solid modeling, defines both the surface and the interior volume and mass (Arabe 2003). A wire frame drawing or surface model may be sufficient for many applications. For photo-realistic renderings, or animations that provide a virtual "walk-through" of a building, the more sophisticated solid modeling software is required. The more complex CAD models require more time and manpower to prepare. Therefore, it is necessary to consider how the 3D model will be employed. A wire frame drawing may be suitable for internal discussions concerning constructability, whereas a more elaborate model would be suitable for explaining a facilities appearance to an owner or potential client.

Many different programs are available to produce 3D visualizations of a construction project. Software companies like AutoDesk and Bentley provide a broad array of programs that are capable of producing 3D images of a building or project. To provide an idea of the various types of software available to produce 3D drawings, the software offerings from several companies are summarized:

AutoDesk

AutoDesk produces several programs that are useful for 3D modeling of buildings and infrastructure designs. These include:

- AutoCAD 2006. AutoCAD is the basic AutoDesk product for 2D and 3D drafting. It is commonly used by engineering designers to produce project plans. Most importantly, the standard file types that it produces are compatible with many other programs. AutoDesk now has a file type called Drawing Web Format (DWF) that allows for improved exchange of information over the Internet. AutoDesk provides a free viewer that allows interested parties, such as the contractor, owner, and subcontractor, to receive and view 3D designs created by the architect or engineer. As we have seen in Chapter 6, it is possible to exchange, review, and mark up design documents in construction web portals.

- AutoDesk Civil 3D. AutoDesk Civil 3D provides three-dimensional renderings of highway, railroad, and channel designs. It allows a dynamic corridor to be created that responds to changes in alignments and profiles, automatically updating project documents. For a construction contractor, the software can calculate complex grading problems and can dynamically balance cut and fill.

- AutoDesk Revit Building. AutoDesk Revit Building is a program that is capable of producing various types of 3D plans and models of a building. It is aimed primarily at architectural firms and includes powerful tools for building information modeling. When design modifications are made Revit Building has the capability of updating model views, drawing sheets, quantity schedules, and sections. 3D drawings constructed in Revit can be displayed in the AutoDesk Buzzsaw web portal.

- AutoDesk Viz. AutoDesk Viz is a program that is capable of producing photorealistic 3D drawings. It is able to accept data from AutoCAD and Revit Building to produce 3D renderings of buildings. The primary use of this capability would be to provide a prototype of how a completed facility will look to an owner or potential client.

Bentley Systems

Bentley Systems has several products that can provide useful 3D visualizations. They include the following:

- Bentley Microstation is widely used for civil engineering design. Microstation has the capability to draw wire frame, surface, and solid models. It also has the capability to produce photorealistic drawings. Microstation is also integrated with Bentley's ProjectWise construction web portal, allowing users to view and modify MicroStation documents while using ProjectWise.

- The GEOPAK Civil Engineering suite works with Microstation to provide 3D models for highway and infrastructure projects. GEOPAK is a software that contains modules for surveying and designing roads, drainage networks, and water and sewer systems. The use of GEOPAK for 3D modeling can allow for the identification of design conflicts earlier in the project cycle. An example of using GEOPAK to identify design conflicts for a highway project is provided in Figure 9-3.

- Bentley Explorer is a 3D model review software that accepts 3D AutoCAD and Microstation models that allows users to perform an interactive walk-through of complex plant models.

- Bentley Explorer Interference Detection is an addon to Bentley Explorer, and it identifies physical interferences. One of the primary applications of this software is the identification of piping interference in industrial plants.

PowerCad

PowerCad has developed 3D software for desktop computers. However, they have also produced software that allows 3D drawings to be viewed in the field using PDA and tablet computers. Their software includes:

- Power CAD Pro 6. This program can be used to produce 3D drawings and it can use files in the standard AutoCAD formats. One of the interesting features that this program has is the ability to create as-built floor plans of a building using a tablet PC. The tablet can be connected to a portable laser-measuring device to record as-built dimensions, and then the as-built plans can be generated on the tablet PC in the field.
- Power CAD CE Pro. This program allows users with PocketPC PDAs to view 3D drawings in the field. The program allows a user to open drawings in DWG and DXF format, and perform markup using pen-based redlining, text callouts, and basic CAD functions. It is also possible to save voice notes. The entire markup is saved in a unique layer of the document that does not affect the original drawing (GiveMePower Corp.).

Case Study of the Application of 3D Modeling to a Complex Project

The following study of the reconstruction of Soldier Field in Chicago illustrates a practical application of 3D CAD to a major construction project. It illustrates some of the major problems encountered when attempting to integrate computer data from different computer systems and programs, and between different organizations. The case study consists of excerpts from an April 14, 2003, article in *ENR* magazine titled "Stadium Engineer Drives Towards 'Paperless' Project" by Nadine M. Post.

Shoehorning a football field and seating bowl, complete with suites, into the horseshoe-shaped perimeter bay of a narrow venue built 80 years ago for track and field was enough to drive many toward distraction. Then, pile on the charge to gut and reconstruct the landmark sports facility in four to six months less than is comfortable and customary for National Football League stadiums. And top that with an unprecedented foray, at least on this scale in the United States, into the largely alien world of 3D computer modeling for design and fabrication of the job's 13,000-ton structural steel frame. It's not surprising that the push toward a "paperless" project at the $365-million makeover of Chicago's Soldier Field caused minor shock waves along the shores of Lake Michigan.

continued

The structural engineer got the green light to create a 3D model and share it with the steel fabricator—for connection detailing and to drive its computer numerically controlled (CNC) fabrication equipment—because the steel structure and its complex radial geometry was on the critical path of the job's 20-month, fast-track schedule. The need to fit the new 61,500-seat bowl snugly into the historic colonnade structure, like a jumbo egg in a small egg cup, also drove the decision to use a 3D approach. "If you can model it in 3-D, you can build it," says Joseph G. Burns, principal in charge for project structural engineer, the Chicago office of Thornton-Tomasetti Engineers.

"For this kind of large, complex project, the 3-D model is the wave of the future," says Joseph Caprile, principal of Lohan Caprile Goettsch, part of project architect, the LW&Z Joint Venture, which also includes Wood & Zapata, Boston. "But the way we develop drawings is going to have to change," says Caprile. "The architect and engineer are going to have to work differently."

On Soldier Field, which could be studied as a guinea pig for 3D modeling on a large, fast-tracked, design-bid-build job, there was "definitely a high learning curve," says Caprile. That mostly emanated from the inexperience of many team members with Xsteel, the project's modeling software produced by Tekla Corp., Espoo, Finland.

"We spent a lot of time" up front "in coordination of the steel and architecture,"

Caprile says. It was "intense," he adds, but once done, the pieces fit together. In the end, the steel portion of the job was conducted in a hybrid manner. The contract documents were issued as traditional 2D drawings with an Xsteel model. "The 3-D model was used as a means of communication between the design and construction team from design development onward," says Burns.

The architect designed the skin in two dimensions. The supplier then created a 3-D model using CATIA software and used the model to drive fabrication equipment. "When there were problems with the 2-D version, we sent 3-D models to the architect," which were then coordinated with the 2D version, says Michael Budd, a vice president of the skin supplier, Permasteelisa/Gartner, Mendota Heights, Minn. Permasteelisa/Gartner was able to pull the Xsteel structural wire frame models into its 3D model via a neutral file format and perform document coordination. "The Xsteel model was helpful," says Budd, as long as it was kept up to date. Budd says his portion of the job got started late because of difficulties getting the 3D geometry from the architect. He doesn't think the delay will affect the end-date. To date, project-wide, there are 48 change orders and more than 4,064 RFIs.

With the project roughly 75% complete, the big push is on. To make the schedule, Hoffman says the developer expects to spend $25 million each month until the end. "We're going to be right on budget," she says.

At Soldier Field, the awkward but giant steps toward a paperless project were driven by the structural engineer. Sources agree that to get the most value from 3D modeling, the effort must be driven from the top down. And to make the process work best, Budd and Dickerson agree that the steel and skin suppliers should be brought into the design process earlier than is customary. "Even if the bid package had been awarded two to three months earlier, it would have made a big difference," says Budd, in terms of avoiding coordination problems. Budd thinks it is going to be a long time before 3D models completely replace 2D drawings. Others disagree, predicting anywhere from five to 10 years until smart models are so pervasive that 2D drawings will be a dim memory.

■ 4D CAD for Construction Planning and Scheduling

One of the most exciting developments in construction information technology is the emergence of commercially available 4D and 5D CAD software. The software has the potential to revolutionize how planning and scheduling of construction is performed. The software we discuss in this section links the construction schedule to a 3D model of a construction project. Some of the software we consider already has the capability to allow users to modify the 3D design, and can automatically reflect the design change in the linked scheduling software and in the 2D design drawings and documents. A 5D model allows changes in the project 3D model to be automatically reflected in the project estimate. The new models are exciting because they not only provide a 3D visualization, but also completely integrate all of the disparate construction software applications and integrate them into a complete building model. In addition, by allowing a visualization of how the project evolves over time, they can revolutionize the planning of projects by construction companies.

A Summary of 4D and 5D CAD Benefits

The benefits of using 4D and 5D CAD are potentially huge. They include:

- Allowance of project participants to understand the impacts of design and construction decisions. A 4D CAD model can allow for improved project understanding and promote collaboration between the owner, designer, and contractor.
- Enhanced constructability analysis of a project.
- Consideration of scheduling conflicts that would not be obvious using only a CPM schedule. It may be difficult to visualize conflicts between crafts at a particular location from a CPM schedule alone.
- Integration of various types of construction software systems like 3D CAD, scheduling, and estimating into a single model.
- Usefulness throughout the entire project cycle from design and project preplanning through the completion of the project. The designer can experiment with various designs and their impact on project cost. During construction, contractors can assess the constructability of different schedule sequences.

When is It Appropriate to Use 4D CAD?

There are some barriers to applying 4D CAD in construction. Some factors to consider include:

- Using 4D CAD requires a high level of training to construct the building model, and to input schedule data. To implement 4D CAD successfully, employees must be properly trained.
- The cost of 4D CAD software is greater than the cost of traditional 2D and 3D models. Consideration must be given to the type of project and its level of complexity. Many projects may not need a 4D model, but could benefit from 3D drawings. Many simple projects are adequately described using 2D drawings.

A construction company must already possess considerable computer expertise to incorporate the new models into its practice. In Chapter 1 we discussed the level of informatization of construction companies. Surely, the early adopters of this technology will be companies that already are comfortable in using advanced computer applications. However, the 4D technology will continue to evolve and will become cheaper and easier to use. Therefore, all construction companies, even if they are not ready to embrace the technology, should follow developments in this area.

Some 4D and 5D CAD Software

There are several different computer programs that have 4D CAD features. We will also discuss programs that have 5D CAD features. The state of the art of this technology is changing rapidly. New products are continuing to arrive on the market. A good source of information about emerging trends in this area is *Constructech* magazine (www.constructech.com) and an online newsletter called CADalyst (www.cadalyst.com/cadalyst/).

A brief summary of some 4D software programs are given below:

- Bentley Navigator. Bentley Navigator is a program with the capability of linking 3D models with either Primavera or Microsoft Project. The software generates simulations of construction schedules and heavy lifts, providing an understanding of object motions and potential clashes in an area of activity (Sheppard 2004). The schedule data is dynamically linked to the 3D drawings, and modifications to the schedule can be immediately visualized in the 3D environment.
- Balfour Technologies fourDscape. Balfour Technologies produces a visualization modeling system called fourDscape. This software is server based and users access the program through a web browser. The software allows users to interact with 3D models of time-dependent information. Users can step forward and backward in time to identify relationships between visualized 3D objects (Sheppard 2004). A principal benefit of fourDscape is that users in various locations can view the 4D visualizations using a web browser. Using the fourDscape program, dates are assigned to 3D objects. The fourDscape model does not include any sort of scheduling algorithm. Therefore, it is necessary to carry out a CPM analysis to determine the dates to assign to the objects. However, the primary benefit of using this program is the capability to allow team members in different locations to view 4D visualizations over the web (Heesom and Mahdjoubi 2004).
- Gehry Technologies Digital Project. Gehry Technologies sells a program called Digital Project (Gehry Technologies 2005). Digital Project is marketed as an advanced building information modeling and construction management system. Gehry Technologies has extensive experience in applying its software to the projects of noted architect Frank Gehry. Gehry's projects are known for their unusual designs and use of curved shapes, and are ideal candidates to be managed using 4D construction. Figure 9-4 shows a 3D visualization produced for a complex building project. The Digital Project software is

based on the CATIA software developed by Dassault Systemes, a French aerospace company. Some interesting features of Digital Project include:

- Bi-directional links between Digital Project and Microsoft Excel.
- Can accept existing 2D and 3D file formats such as DWG and DXF.
- Project management functions can be integrated with the 4D-building model. Using Microsoft Office, linked templates for design issues, change orders, and requests for information can be produced. Figure 9-5 illustrates the generation process that allows quantities and shop drawings to be produced from the 3D drawings. Figure 9-6 shows the 3D model for a high-rise building along with a window showing the extracted quantities. Figure 9-7 shows how Requests for Information can be linked to 3D drawings.
- 4D navigation is provided by linking the 3D model to Primavera. Figure 9-8 shows how the construction sequence can be simulated using Primavera.

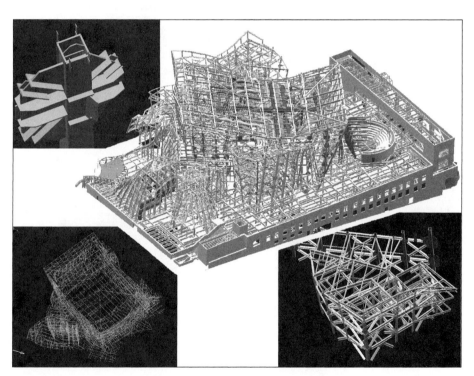

FIGURE 9-4 3D Visualizations from Digital Project
Courtesy of: Gehrey Technologies, Inc.

3D, 4D, and 5D CAD Applications in Construction

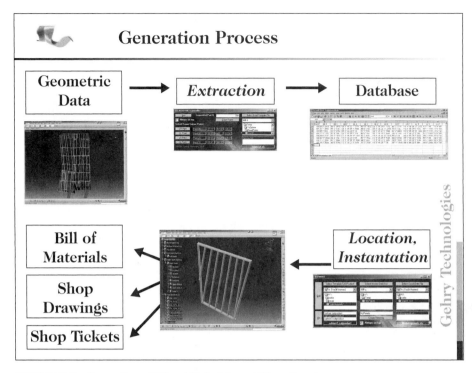

FIGURE 9-5 Generation of Bills of Materials and Shop Drawings
Courtesy of: Gehrey Technologies, Inc.

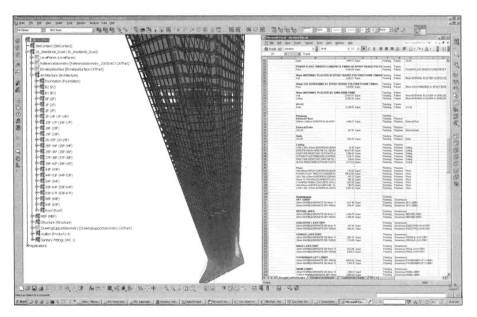

FIGURE 9-6 Bill of Quantities Extraction
Courtesy of: Gehrey Technologies, Inc.

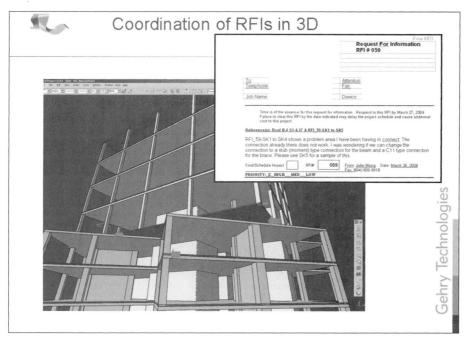

FIGURE 9-7 Coordination of RFIs in 3D
Courtesy of: Gehrey Technologies, Inc.

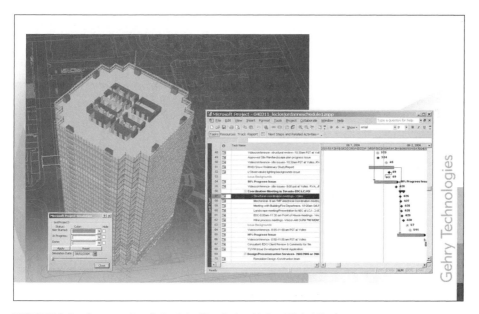

FIGURE 9-8 Construction Schedule Simulation Using Digital Project
Courtesy of: Gehrey Technologies, Inc.

3D, 4D, and 5D CAD Applications in Construction

- Common Point 4D. Common Point 4D is another powerful 4D CAD program that can be used for both constructability analysis of a structure and the layout and planning of a construction site. The program can accept scheduling data from both Primavera and Microsoft Project. It also accepts various types of CAD files as input. One interesting feature is the ability to accept a Tekkla Structures LE file, which is a program, used to create 3D drawings of steel detailing.

There are several 4D CAD models that focus on plant construction and maintenance. These programs are useful, not just for the original construction, but are designed to be used throughout the lifecycle of the plant to handle maintenance and retrofits. These programs serve the additional role of a database of the plants components and layout. These programs include:

- Intergraph SmartPlant Review
- Computer Systems Associates Plant CMS/Database

This review of the software is not intended to be comprehensive. New software is continuously being developed. This review of some of the best-known software indicates that there are differences in approach and features among the different programs. 4D software must be carefully evaluated before purchase to ensure that its features are the best match for the types of projects a firm conducts.

5D CAD Software: Graphisoft Constructor

Graphisoft Constructor is a 5D model that provides an integrated solution for project visualization, scheduling, and estimating. The basic Constructor program provides 4D modeling using integration with Primavera. Figure 9-9 shows a 3D model produced by the Constructor program. This figure shows a cross-section of the California Academy of Science in San Francisco. The building is very unusual and has many features that are difficult to describe in two dimensions. The ability to view changes in the project over time in 3D can serve as a valuable planning and collaboration tool. Figure 9-10 shows a report generated by Constructor that identifies design conflicts and omissions in the design.

158

Chapter 9

FIGURE 9-9 3D Cross Section Produced by Graphisoft Constructor
Courtesy of: Graphisoft and California Academy of Science, Webcor Builders

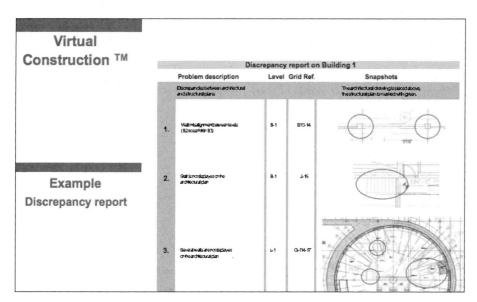

FIGURE 9-10 Report Showing Design Conflicts and Omissions
Courtesy of: Graphisoft

3D, 4D, and 5D CAD Applications in Construction

Additional modules of the Constructor program are Estimator and Control. Figure 9-11 shows the various ways the Constructor program can be employed to enhance constructability on a project. When using Estimator, every object in the 3D model is connected to an estimating recipe. Figure 9-12 shows how a 3D object, in this case a column, is linked to a recipe in Estimator where the materials, construction methods, and resources are described. In turn, this recipe information can be used to define schedule durations. The Estimator program includes the capability to perform Monte Carlo simulations if the user inputs low, most likely, and high costs for an estimate item. Cost variance reports can be produced that show project elements that may be prone to cost increases. The Estimator program is linked to the 3D building model and can also be used to rapidly assess the cost impacts of design alternatives. Figure 9-13 shows how two different types of floor slab systems can be evaluated. Changes in the 3D model are automatically reflected in the estimate.

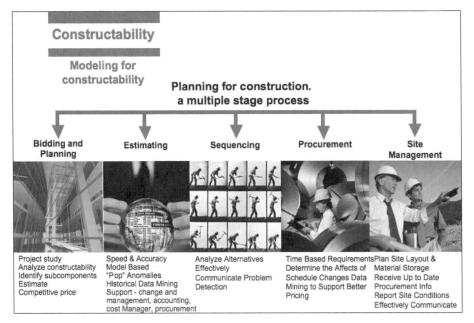

FIGURE 9-11 Ways to Employ the Constructor Program for Constructability Analysis
Courtesy of: Graphisoft

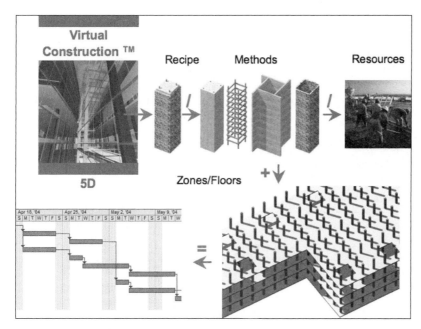

FIGURE 9-12 Estimator Recipes Linked to 3D CAD
Courtesy of: Graphisoft, Patent Pending

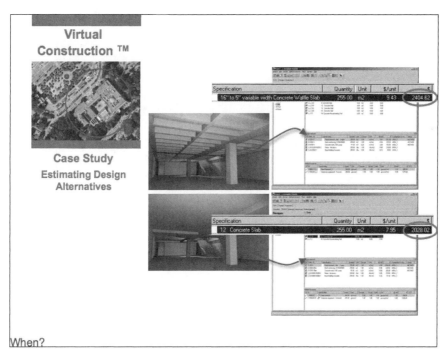

FIGURE 9-13 Evaluation of Design Alternatives
Courtesy of: Graphisoft

3D, 4D, and 5D CAD Applications in Construction

The Control software is an interesting feature of the Graphisoft Constructor offerings because it provides an alternative paradigm for scheduling based on location. The Control software generates a line-of-balance diagram for a selected location in a building. The line-of-balance diagram explicitly shows where conflicts occur between different resources at the selected location. The line-of-balance schedules are created from Estimator recipes and 3D model quantity information. Tasks can be defined by grouping materials and defining a production factor. Figure 9-14 shows how a line-of-balance diagram created in Control is linked to a 4D view of the project. Created schedules can be imported into the Constructor 5D model and simulated using the Control software. Design updates can be reflected in the Control schedule by re-importing the material quantities from the Estimator software (Gallello et al.).

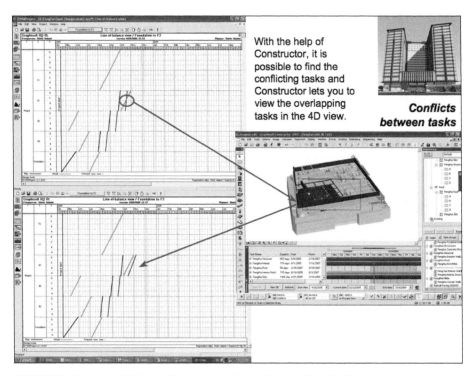

FIGURE 9-14   Line-of-Balance Diagrams to Detect Scheduling Conflicts
Courtesy of: Graphisoft

■ Software Web Sites

See the following table for links to the software discussed in this chapter.

Software	Web Site
AutoDesk Software	http://usa.autodesk.com/
Bentley Systems Software	http://www.bentley.com/en-us/
PowerCAD	http://www.givemepower.com/
Balfour Technologies fourDscape	http://www.bal4.com
Gehry Technologies Digital Project	http://www.gehrytechnologies.com
Common Point 4D	http://www.commonpointinc.com
Intergraph SmartPlant Review	http://ppm.intergraph.com/
Computer Systems Associates Plant CMS/ Database	http://www.csaatl.com/
Graphisoft Constructor	http://www.graphisoft.com/
Tekla Structures	http://www.tekla.com/

■ Conclusions

4D CAD and 5D CAD are emerging as a new way to manage projects. 4D CAD has the potential to reduce construction costs by allowing for better planning and constructability analysis, reducing the need for rework. The major barriers to implementing 4D CAD are the additional time and cost required to develop a 4D model for a project, the cost of the 4D software and the need to have highly trained, computer-literate staff. 4D CAD software will become easier to use and less costly over time. This will enable an increasing number of construction contractors to employ it on their projects.

■ References

Arabe, K. 2003. The Basics of CAD & CAM: Industrial Market Trends. *ThomasNet Industrial News Room.* http://news.thomasnet.com/IMT/archives/2003/05/the_basics_of_c_2.html (Accessed December 5, 2005).

Cory, C. 2001. Utilization of 2D, 3D, or 4D CAD in Construction Communication Documentation. *Fifth International Conference on Information Visualisation* 219–224.

Gallelo, D., Broekmaat, M., and Freeman, C. Virtual Construction Benefits. http://www.graphisoft.com/products/construction/white_papers/whitepaper2.html (Accessed December 1, 2005).

Gehry Technologies. 2005. Gehry Technologies Announces Major New Software Update. http://www.gehrytechnologies.com/company-press-07-15-2005.html (Accessed December 10, 2005).

GiveMePower Corp. PowerCAD Mobile Handheld Solutions. http://www.givemepower.com/products/mobile.cfm (Accessed December 9, 2005).

Heesom, D. and Mahdjoubi, L. 2004. Trends of 4D CAD applications for construction planning. *Construction Management and Economics.* 22:171–182.

Sheppard, L. 2004. Virtual Building for Construction Projects. *Computer Graphics and Applications.* 24:6–12.

CHAPTER 10

Software for Construction Accounting and Project Cost Control

INTRODUCTION

The financial management of a construction company can be complex. Even the smallest companies can benefit from using computers to manage their projects. As we will discuss, there is inexpensive software aimed at small contractors. Large construction companies have complex requirements to obtain information about the financial performance of a large portfolio of projects and to monitor the financial condition of the firm. Computers are widely used as an aid in construction accounting and project cost control. Construction is a unique industry, with work done on a project basis, using many subcontractors. Specialized accounting software is required to accommodate the specialized nature of the construction industry.

There is a wide variation in the size of firms participating in the construction industry, and their accounting requirements are diverse. Therefore, accounting software is available for only a few hundred dollars that is tailored to homebuilders and small contractors.

On the other end of the spectrum, large contractors may use very expensive networked versions of accounting software tailored to their needs. In addition, sophisticated computer users in the construction industry are beginning to see the importance of integration and are linking their accounting systems with estimating software, project control systems, and project schedules to receive more timely information about their firm's financial health and project cost performance.

◼ Purpose and Function of Accounting and Cost Control Software

Construction accounting software typically can fulfill two distinct management goals. These are:

- The ability to determine the profitability of individual projects and control costs on individual projects.
- The ability to assess the entire company's financial situation, its profitability, and the production of required financial statements like the balance sheet and income statement.

Figure 10-1 shows the relationship between the two aims. To perform either function, data is collected from the field. Most construction accounting software allows the field data to be entered once, and then the data can be used for both accounting and cost control purposes. The cost control system typically accepts the project estimate to establish the budget for a project. The objective of the cost control system is to establish the profitability of an individual project, and to identify line items in the project where cost deviations are occurring, so corrective action can be taken to avoid losses on the project. The accounting functions provide management with a way of assessing the firm's financial health on an aggregate basis and to manage the firm's cash flow. Many different functions can be performed by construction accounting software. Construction companies typically require the following types of financial management processes (Peterson 2005):

- Preparation of the basic financial documents, such as the balance sheet and the income statement.
- Computer software capable of maintaining an appropriate accounting system including accounts recievable and accounts payable.
- Projecting the cost at completion of individual projects.
- Calculating if projects are over- or under-billed. Because construction is done on a project basis, and a project may not be complete at the end of an accounting period, it is important to calculate over- and under-billing to ensure correct reporting of financial results.
- Controlling the costs of individual projects.
- Analyzing the profitability of different projects.
- Developing overhead budgets and tracking them over time.

- Analyzing the profitability of different parts of a company.
- Monitoring equipment costs and depreciation.
- Managing a construction company's cash flow to ensure that it can take on new business and meet current obligations.

There are many different construction accounting software packages available. The following section provides an example of the capabilities of construction accounting software by looking at an example of the JobView software in depth. Later in the chapter, some of the most popular accounting programs will be discussed.

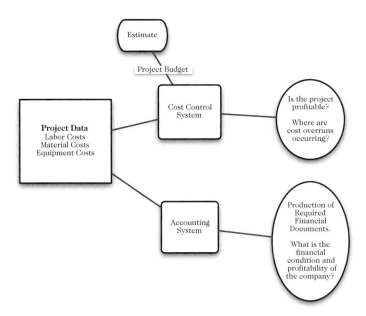

FIGURE 10-1 The Relationship between Accounting and Cost Control

◼ Some Accounting and Cost Control Examples Using JobView

To provide an introduction to the capabilities of construction accounting and cost control software, we will discuss the capabilities of the JobView software by A-Systems. JobView is intended for small- and medium-sized contractors, yet it still has many different features and can be run on a local area network to allow multiple users to access data.

Some of the important features of the JobView software include:

- Production of financial statements such as the income statement and balance sheet. Figure 10-2 shows a balance sheet and Figure 10-3 shows an income statement produced by JobView. The income statement defines the company's profits or loss for a given time period. The balance sheet provides a snapshot of the company's financial situation for a particular date.

- Over- and under-billing can be calculated by the accounting system. A unique feature of the construction industry is that a construction project often occurs over several accounting periods. A contractor may have over- or under-billed the actual value earned on individual projects. The JobView software allows the situation to be assessed for its impact on income recognition in the accounting system. This is shown in Figure 10-4.
- Various features for project cost control are included in the software. Figure 10-5 shows an overview of project cost and profit. Figure 10-6 shows a text report produced by the program that summarizes job costs by cost codes.
- The software is capable of tracking equipment costs to measure the profitability of equipment use. Figure 10-7 shows a report that tracks the costs associated with a construction company's equipment.
- Another unique feature of the construction industry is the widespread use of subcontractors. Figure 10-8 shows a report that tracks payments to the many subcontractors used on a project.
- Naturally, the usefulness of the accounting system depends on the information that is input. Therefore, it is necessary to provide input screens that allow equipment, labor, and material costs to be collected. Figure 10-9 shows how a purchase order for materials is input to the accounting system.

This review of JobView illustrates the basic capabilities of construction accounting software. The strength of the JobView software is that it provides many features, yet is relatively easy for a small- or medium-sized contractor to learn and assimilate.

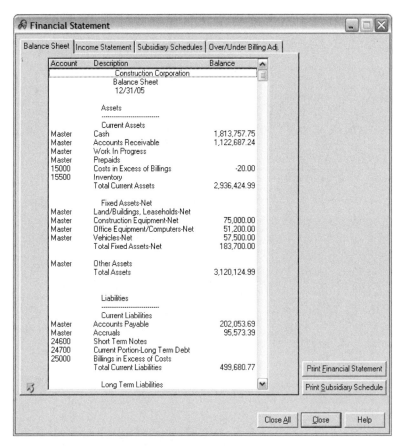

FIGURE 10-2 JobView Balance Sheet

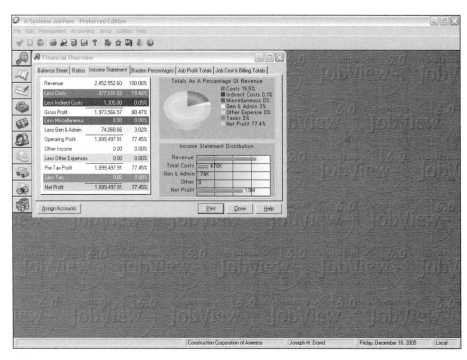

FIGURE 10-3   Income Statement

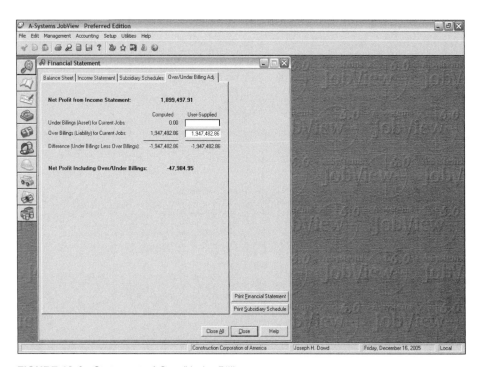

FIGURE 10-4   Statement of Over/Under Billing

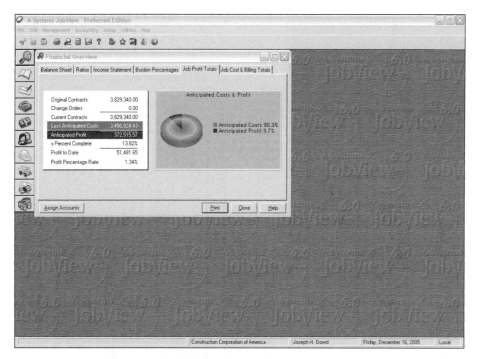

FIGURE 10-5 Overview of Project Costs and Profit

Construction Corporation of America
Complete Job Cost Analysis

12/16/05
10:16AM

Job 100

Cost Code	Description	Quantity Estimate	To-Date	% Complete Report Calc	Budget Original	Current	Remaining	Current Variance	Actual Cost To-Date	Projected Cost	Projected Overrun
100	**A-Systems Corporate Office**			Contract:	600 000.00	600 000.00					
01000 G	General Requirements		0.0	0	10 109.00	10 109.00	10 109.00	0.00	0.00	10 109.00	0.00
01000 L	General Requirements	23.0	69.0 Hours	100	0.00	0.00	-826.73	826.73	826.73	826.73	826.73
01011 L	Supervision	258.0	400.0 Hours	100	8 500.00	8 500.00	4 106.00	-4 106.00	4 394.00	4 394.00	-4 106.00
01020 G	Overhead		0.0	0	25 777.00	25 777.00	25 777.00	0.00	0.00	25 777.00	0.00
01030 G	Profit as Bid		0.0	0	26 479.00	26 479.00	26 479.00	0.00	0.00	26 479.00	0.00
01515 S	Project Signs	Marsh; Henry B.		0	500.00	500.00	500.00	0.00	0.00	500.00	0.00
01607 G	Building Survey		0.0	0	1 500.00	1 500.00	1 500.00	0.00	0.00	1 500.00	0.00
01609 G	Subsistence		0.0	0	60 630.00	60 630.00	60 630.00	0.00	0.00	60 630.00	0.00
General Requirements	Subtotal:	261.0	469.0 Hours	4	133 495.00	133 495.00	128 274.27	-3 279.27	5 220.73	130 215.73	-3 279.27
02222 S	Structure Excavation	Life-Scape		46	46 540.00	46 510.00	25 010.00	0.00	21 500.00	46 510.00	0.00
02801 S	Landscaping	General Tree Services, Inc.		87	17 280.00	17 280.00	2 276.25	0.00	15 003.75	17 280.00	0.00
Sitework	Subtotal:		0.0 Hours	57	63 820.00	63 790.00	27 286.25	0.00	36 503.75	63 790.00	0.00
03201 S	Rebar Reinforcing	Steiner Corporation		63	4 787.00	4 787.00	1 787.00	0.00	3 000.00	4 787.00	0.00
03301 G	Pour Footings		0.0	100	1 000.00	1 000.00	-542.69	542.69	1 542.69	1 542.69	542.69
03301 L	Pour Footings		16.0 Hours	15	1 000.00	1 000.00	854.78	0.00	145.22	1 000.00	0.00
03301 M	Pour Footings		0.0	0	40 000.00	40 000.00	40 000.00	0.00	0.00	40 000.00	0.00
03301 S	Pour Footings	Pacific Rim Constructors		100	38 000.00	38 000.00	0.00	0.00	38 000.00	38 000.00	0.00
03311 G	Pour Ground Slabs		0.0	0	1 600.00	1 600.00	1 600.00	0.00	0.00	1 600.00	0.00
03311 L	Pour Ground Slabs	33.0	0.0 Hours	0	1 000.00	1 000.00	1 000.00	0.00	0.00	1 000.00	0.00
03311 M	Pour Ground Slabs		0.0	0	5 700.00	5 700.00	5 700.00	0.00	0.00	5 700.00	0.00
03311 S	Pour Ground Slabs	Biltmore Contracting		0	2 600.00	2 600.00	2 600.00	0.00	0.00	2 600.00	0.00
Concrete	Subtotal:	33.0	16.0 Hours	44	95 687.00	95 687.00	52 999.09	542.69	42 687.91	96 229.69	542.69
04210 S	Brick Masonry	Mason's Supply Company		0	37 250.00	37 000.00	37 000.00	0.00	0.00	37 000.00	0.00
04250 S	Ceramic Veneer	Oster; Robert		0	6 495.00	6 495.00	6 495.00	0.00	0.00	6 495.00	0.00

FIGURE 10-6 Job Costs by Cost Code

Construction Corporation of America
Equipment Analysis

ID	Description	VIN	License	Month to Date Cost	Year to Date Cost	Cumulative Cost
BH-1	Backhoe #1	09872U34ROIJW	387HE	625.00	625.00	625.00
CAT-1	Caterpillar D-9	C4580-2345	0065888	30.00	30.00	30.00
CAT-2	'Bobcat' Small Dozer	079-WW11021-94	UFF-093	300.00	300.00	300.00
CMP1	Pneumatic Drill Compressor			0.00	0.00	0.00
CMP2	Air Compressor			0.00	0.00	0.00
DW-2	Ditch Witch (2)	722-AWZ-92001	NRO-551	0.00	0.00	0.00
TR-1	Ford Pickup	09U123409135	AS4056	150.00	150.00	150.00
TR-2	Flat Bed Truck	772-318910-QA	WNJ-0201	200.00	200.00	200.00
TR-3	Chevy Pickup (Red)	NHH-027001-92	HAL-9000	0.00	0.00	0.00
TR-4	Chevy Pickup (Blue)	NHH-0034509-21	RKW-516	0.00	0.00	0.00
TRUCK	Ford 1/2 Ton Pickup	9821734 (VIN)	327RL	0.00	0.00	0.00
	Totals:			1 305.00	1 305.00	1 305.00

FIGURE 10-7   Equipment Costs Report

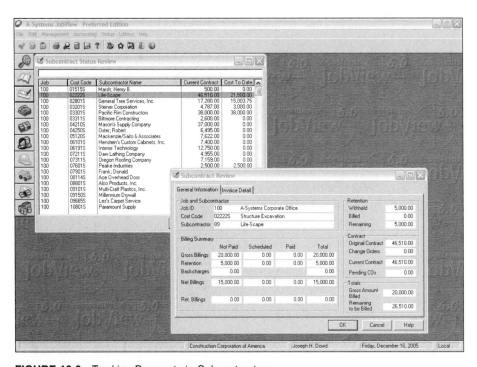

FIGURE 10-8   Tracking Payments to Subcontractors

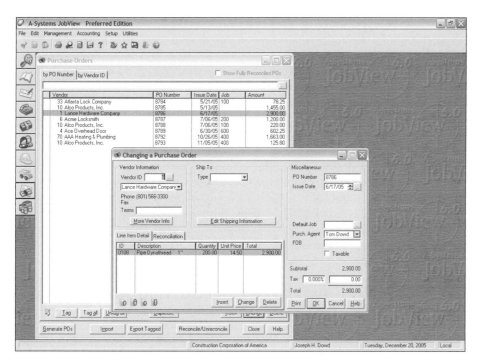

FIGURE 10-9 Purchase Order Input Form

■ A Consideration of More Complex Accounting Software: Dexter & Chaney's Forefront

An example of a highly sophisticated accounting application is the Dexter & Chaney Forefront software. The Forefront software can be specialized for use in general construction, heavy construction, electrical construction, mechanical construction, and utility construction. The Forefront system is based on a series of modules. A user is required to purchase a basic set of modules and can then select additional modules to meet the particular requirements of its business. The software is client-server software and is designed to be installed on a company's network. Wide-area networking technology allows users at diverse geographic locations to access the accounting and cost control system.

For example, the base system required by a Heavy Contractor would include Accounts Payable, Accounts Recievable, Cash Management, and General Ledger software. A user could then also select job costing, equipment tracking, payroll, human resources, and project management software. The project management feature incorporates many of the features of a web portal that we discussed in Chapter 6. In effect, the Dexter & Chaney software, through the addition of the construction management module, can link and integrate a company's accounting system to project management systems that handle the day-to-day interchange of project documents.

One of the interesting features of the Dexter & Chaney software is the integration of document imaging with the software. Using document imaging, all project paperwork including handwritten forms can be scanned into the accounting system as TIFF files. Document imaging is integrated with many of the Forefront modules, including Accounts Payable, Accounts Receivable, Equipment Control, Equipment Tracking, General Ledger, Human Resources, Job Cost, Payroll, Project Management, Purchase Order, and Work Order.

Scanned documents are automatically indexed to allow for easy retrieval. The software allows users to build CDs of related documents. A good example of the use of this feature is the ability to put together all of the documents for a time and material billing, including timesheets and material invoices for transmission to a client. Rather than the need to file and make multiple copies of large volumes of paper a CD can be created instead. Most scanners are suitable for the job. Although almost any scanner can produce suitable files, a commercial scanner with the capability of scanning both sides of a page and high scanning speeds (around 15 pages per minute) is probably most appropriate for a construction company. They cost around $800 (Dexter & Chaney).

■ A Review of Some Available Construction Accounting Software

There are many different construction accounting programs available. They range from simple programs meant to run on a single computer to large, network-capable systems that can be used for cost control on multiple complex projects. In this section, some of the most popular accounting software will be discussed. These programs include:

- QuickBooks. Intuit, the well-known maker of programs like Quicken, offer an entry-level construction accounting software called QuickBooks: Premier Contractor Edition 2006, starting at around $400 for a single user. This software is useful for homebuilders, remodelers, specialty contractors, and subcontractors.
- Master Builder. Master Builder is accounting and cost control software that incorporates estimating, scheduling, and document management functions for use by small- and medium-sized contractors.
- Peachtree Premium Accounting for Construction. Peachtree Premium Accounting for Construction by Sage is a version of the well-known Peachtree accounting software tailored for the construction industry. Sage is also the developer of the more advanced Timberline software.
- StarBuilder and StarProject. StarBuilder is construction accounting and job cost software. StarProject is reporting and document management software to communicate project financial performance to managers. The software developer, Geac, sells versions tailored to various segments of the construction industry. StarBuilder is built using Microsoft SQL server (a database for use with server-based programs) and allows import and export of data between the accounting program and its other modules and programs such as Microsoft Word and Excel.

- The Profit Builder. The Profit Builder produces accounting software for homebuilders. The software comes in several versions for commercial builders, plan homebuilders, and one-of-a-kind homebuilders.
- Sage Timberline Office. Timberline office is a powerful program for project cost management, accounting, estimating, and project management. We have already discussed the estimating capabilities of the software in Chapter 3. This software is widely used by commercial and industrial building contractors and provides complete integration between the accounting system, cost control, scheduling, and project management.
- The Maxwell Management Suite. This is another high-end program that integrates accounting, cost control, and document management. An interesting capability of this program is the ability to use a web site called MaxCentral (hosted by the software developer Maxwell Systems) to allow financial data such as equipment usage and timesheets to be collected in the field. MaxCentral allows users to distribute information from their accounting systems: share and access Microsoft Word files, Excel spreadsheets, CAD files, and job pictures (Maxwell Systems). Maxwell Systems also sells a program called American Contractor that offers accounting, cost control, and document management features.
- Streetsmarts. Cheetah Streetsmarts is accounting, cost control, and project management software intended for heavy construction. Construction material suppliers can also use the software.

This is only a partial listing of the software that is available. This listing has progressed through simple software for a small contractor to highly sophisticated software for contractors building multiple complex projects. In addition, the listing illustrates that the software offerings are tailored to different types of construction, such as homebuilders, heavy contractors, and commercial contractors. With starting prices around $400, computerized accounting is accessible to all construction contractors. A listing of web sites for construction software providers is given at the end of the chapter.

■ The Trend Toward Software Integration

Profitability is naturally the major concern of any business endeavor. Therefore, it is natural for a construction contractor to seek management tools that provide both top executives and managers in the field with improved and timely information about project financial performance. This market force has led software developers to increasingly develop accounting and project cost control programs that are integrated with estimating, scheduling, and project management. The high-end estimating software has increasingly moved in this direction, as the Dexter & Chaney Forefront and Sage Timberline software illustrates.

Figure 10-10 illustrates the possible interconnections between construction software and a continued convergence of accounting, web portal software for managing project financial documents, estimating, and scheduling software into powerful integrated solutions. It should be noted that software that integrates a firm's accounting with project document management is typically implemented on private client-server networks for the security of the firm's financial data. Many of the web portals we have discussed provide Internet access that is possibly less secure.

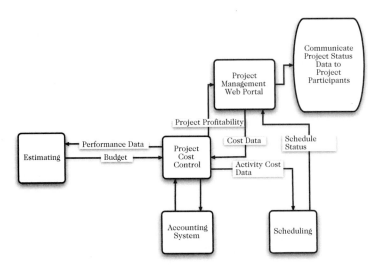

FIGURE 10-10 Potential Connections Between Construction Software Systems

Conclusions

The available software for construction accounting and project cost control varies from basic programs that run on a single personal computer to complex systems capable of accommodating many users on a wide area network. All programs offer the basic functions of a general ledger, accounts payable, accounts receivable, and generation of financial reports and documents. Most programs also provide some form of project cost control and the ability to measure profitability on individual projects. As the software becomes more expensive, additional features are added, particularly the capability to integrate document management, estimating, and scheduling with the accounting software.

Links to Accounting Software Web Pages

Software	Web link
JobView	http://www.a-systems.net
Forefront	http://www.dexterchaney.com/
Quickbooks	http://www.intuit.com/
MasterBuilder	https://masterbuilder.intuit.com/
Peachtree	http://www.peachtree.com/
StarBuilder	http://www.construction.geac.com/
The Profit Builder	http://www.cdci.com/
Timberline Office	http://www.sagetimberlineoffice.com/
Maxwell Management Suite	http://www.maxwellmanagementsuite.com/index.html
American Contractor	http://www.amercon.com/
Cheetah Streetsmarts	http://www.cheetahware.com/

References

Dexter & Chaney. Document Imaging. http://www.dexterchaney.com/document-imaging-software/dc/m2326t25s7r13p7.html (Accessed December 21, 2005).

Maxwell Systems, Inc. MaxCentral. https://www.maxcentral.net/MaxCentral/about/about.asp (Accessed December 21, 2005).

Peterson, S. 2004. *Construction Accounting and Financial Management*. Upper Saddle River, NJ: Prentice Hall.

CHAPTER 11

Construction Applications of Mobile and Wireless Computing

INTRODUCTION

The use of wireless computing in the United States is growing rapidly. It was estimated in 2004 that 50% of U.S. businesses already employ wireless networks in some capacity (Sandsmark 2004). Computer manufacturers are increasingly including wireless equipment as standard items in their laptops. Coupled with the increasing portability of computers and PDAs, the use of computers in the field of construction is increasing rapidly. In this chapter we will explore how traditional applications, such as scheduling and estimating, can be enhanced using mobile computers. In addition, we will explore how Internet access in the field opens up a broad range of applications for use at the construction work force.

■ Computers in the Field

Construction companies commonly use personal computers in their field offices. A survey of the use of wireless computing in construction (Williams et al. 2006) indicates that construction companies already provide personal computers and web access in their project trailers. The survey of 58 managers in the construction industry indicated that 95% of the firms surveyed used computers at the construction site and 91% of the firms surveyed provided web access at construction sites. Surely the need to communicate using e-mail and the proliferation of web portal software described in Chapter 6 has made Internet access a requirement for many project offices. The study also found that around 35% of respondents said they were using wireless networks on their projects, and indicated that almost all of the respondents were interested in eventually using wireless technology. In this chapter, we will explore the existing uses of the computer at the construction site and the expanding possibilities for using mobile and wireless computing at the construction work force. We will consider programs for managing the field office, stand-alone uses of mobile computers, integration of the use of mobile computers with existing construction software, and the use of wireless computing to provide network access to users in the field. The initial construction applications of mobile computers focused on collecting data in the field. Typically there was no wireless access. Field personnel typically attached their mobile device to a personal computer in the field office. Now, wireless networking allows computers at the construction site to be networked together and to have Internet access. This allows users in the field to access all types of construction software and to input data into web portal and knowledge management systems instantly. It also provides enhanced knowledge to constructors in the field through web-based examples of how to conduct construction operations.

■ Benefits of Mobile Computing

One of the primary benefits of mobile computing is the ability to reduce the need for field personnel to return to the project office for information and documentation. Additionally, data and information can be entered into a company's computer systems more rapidly so managers can quickly spot where problems are occurring. Laing-O'Rourke, a large British contractor, has estimated that widespread implementation of wireless computers on its project will boost productivity 20% to 30% (Intel). Several beneficial uses of portable computers and wireless technologies will be discussed in this chapter. They include:

- Providing construction knowledge in the field through electronic books and web access.
- Access to traditional construction applications in the field, such as schedule and estimate data. Users cannot only view estimates and schedules, they can also modify the data based on their observations in the field.
- Wireless access to web portals and content management systems. As the use of web-based systems is increasing in the construction industry, the ability to access the web at the construction site to view project documents is becoming increasingly important.
- A mobile computer to provide access to project design information and quality standards can replace specifications and plans.

Hardware for Mobile Computing

There are several types of devices that can be employed in the field for mobile computing:

- Laptop computers, with their increasing affordability, are being increasingly used in the field.
- Tablet computers can accept input via a touch-sensitive screen and can accept handwriting and provide an efficient way of collecting data using forms and its touch-sensitive capabilities.
- Personnel Digital Assistants (PDAs) are becoming cheaper and more powerful. Their small size and light weight make them ideal for applications in the field.

PDAs offer a reduced set of features compared to full-fledged computers, but their small size and low cost make them attractive. Two popular types of devices are those that run the Pocket PC operating system and devices manufactured by Palm.

All of the mobile computing devices can be purchased in "ruggedized" versions suitable for use in the field. Figure 11-1 shows a ruggedized Pocket PC PDA device manufactured by TDS. Mobile computing devices can provide construction field personnel with a lightweight means of bringing documentation to remote construction activities. Mobile computers can provide access to documentation and manuals that would normally be too heavy or bulky to carry. In addition, it can provide the material in a highly organized electronic format so it can be rapidly accessed during construction activities. The potential exists to employ mobile computers and PDAs widely in construction. Mobile computers can be used to run programs for a wide variety of applications, and can provide e-mail and Internet access.

FIGURE 11-1 A Ruggedized Pocket PC
Courtesy of: TDS Recon by Tripod Data Systems

■ Wireless Networks

A wireless network uses radio frequency technology to transmit and receive data. To implement wireless networks at the construction site requires some understanding of what types of networks are available and their capabilities. The development for wireless networks has many potential applications in the construction industry. Wireless networking, coupled with mobile computers, can provide people in the field with access to most of the programs and web-based software described in this book. The advent of the wireless network frees managers and workers from the need to go to an office to use the computer, and they can spend more time in the field, managing the construction process.

Wi-Fi Networks

The most common type of Wireless Local Area Network (WLAN) is the Wi-Fi network. Wi-Fi stands for Wireless Fidelity. When dealing with Wi-Fi equipment, a common term is 802.11. 802.11 refers to the family of IEEE standards for building Wi-Fi networks. Most Wi-Fi equipment conforms to these standards. The 802.11 standards are evolving over time, with the main standards currently being 802.11a, 802.11b, and 802.11g. 802.11g offer higher speeds than

802.11b. 802.11a networks have a shorter range but are less prone to interference from other electronic devices. Both 802.11g and 802.11b have ranges of about 300 feet, but the range can be extended at a construction project using devices called repeaters (Sandsmark 2004). A new standard, 802.11n, is currently being developed. It will also offer faster data speeds and longer range (Dodd 2005).

Figure 11-2 shows a configuration of how Wi-Fi equipment might be employed on a construction project. Depending on the configuration of the construction site, several access points will probably be needed to provide full coverage of the construction site. The configured network can be a mix of wireless and cabled network, or entirely wireless.

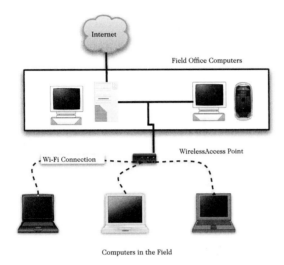

FIGURE 11-2 A Typical Wi-Fi Network

Wireless networking technologies are continuing to evolve. Many cities are beginning to deploy Wi-Fi networks that provide coverage throughout a metropolitan area. These developments indicate that access to Wi-Fi networking for construction contractors will be easier in the future.

Wi-Fi on the Construction Site

Most portable computers and PDAs come installed with Wi-Fi networking cards adhering to the 802.11 standard. To implement a Wi-Fi network, a wireless router is required. To provide Internet access, the router is connected to an incoming high-speed Internet connection and then wirelessly broadcasts to computers equipped with Wi-Fi cards. Typically, a high-speed connection would be brought to the project field office and the router would be installed there. Therefore, Wi-Fi networks to provide an Internet connection for field computers can only be implemented at sites where it is possible to provide a DSL or other high-speed line to the project office.

Examples of Wi-Fi Use in Construction

There are some interesting applications for companies that already use Wi-Fi networking in construction. This section describes two examples.

Wireless Network Applications by Webcor

Webcor Builders is a 975-employee company that specializes in large commercial projects. Webcor has implemented wireless networking to allow project superintendents to spend more time in the field and to provide access to project plans and documentation in the field. Webcor also uses digital photography and wireless PDAs to make it possible to send images of problems in the field to designers to speed the change order process. A tablet PC is used for site inspections and the information is instantly transmitted to Webcor's central database (Greengard 2004). Webcor connects multiple construction projects by a WAN. A high-speed data line is provided for each construction trailer. Then, 802.11g and 802.11b networks provide data access to workers on the construction site. Webcor has also found that it is relatively easy to set up a wireless network in just a few hours. Webcor often extends the range of wireless networks by placing repeaters and signal boosters around the job site.

Wireless Networking by Laing-O'Rourke

Laing-O'Rourke is a large design/construction company in Great Britain. It provides employees in the field with laptop computers and installs wireless hotspots in its offices and on its construction projects. Laing-O'Rourke believes that decision making is faster and that needed design modifications can be made more rapidly, avoiding rework and quality problems, with the use of laptop computers in the field. Laing-O'Rourke estimates that this implementation of wireless computing boosts productivity by 20% to 30% without increasing IT costs (Intel).

Wireless Connectivity Using the Cellular Network

The cellular service providers are now offering high-speed data connections. The cellular phone companies are now providing a WAN for broadband data communications. These data connections provide speeds about three times faster than those of a dial-up connection on a regular phone line. Two widely used services in the United States are Cingular's EDGE network and Verizon's BroadbandAccess. The cellular WAN can be used by a PDA with a built-in phone, by a computer connected to a cell phone, or through a card that fits in the laptop. This mode of access is not as fast as using Wi-Fi but it does not require networking hardware to be installed at the project site. Cingular markets a router that is connected to its EDGE network. This has been used in construction field offices where phone lines are not available, or have not yet been installed. All of the computers in the office are connected to the router that provides wireless access to Cingular's network (Brown 2005).

Like all of the technologies discussed in this book, the wireless services offered by the phone companies are continuing to evolve rapidly. The stages of cellular phone development are often described as generations. When discussing wireless cell services, the terms 2G, 2.5G, and 3G are often heard. Cell phone providers are now in the process of transitioning from 2G and 2.5G to 3G services. Networks like Verizon's BroadbandAccess and

Cingular's Edge are considered 2.5G services. Third-generation wireless services will be implemented by the wireless carriers in the near future and will further increase the data transfer rates of the wireless cellular network.

Virtual Private Networks

A commonly used term when dealing with wireless access for remote users is Virtual Private Networks (VPN). A VPN has several advantages for use in construction and can be employed in several ways. A VPN is an arrangement that allows connections between offices, remote workers, and the Internet without requiring private lines. (See the discussion of WAN in Chapter 1.) A VPN's primary advantage in the context of mobile computing at the construction site is that it allows a user in the field to securely access corporate intranets and corporate files for use in the field. If a construction company has a networked version of estimating or scheduling software on the company's LAN, it would be possible for a user in the field with a laptop to access project data and use it on his or her own personal computer. Second, users in the field can access web-based programs such as a weblog and a content management system that are running on the private corporate intranet (Dodd 2005).

Thin Client Computing

Thin client computing represents the step beyond a VPN, and resembles, in some instances, the original use of terminals with mainframe computers. A thin client is defined as a computer that gets its computing resources from a server (Sinclair and Merkow 2000). The most common software used to provide thin client access to remote computers is provided by Citrix Systems.

Using the Citrix software, computers can access and run programs residing on a server, without the need to install the software on the remote computer. For example, by using a Citrix connection to a server a user could access the Primavera scheduling software without having Primavera installed on his or her own computer. This contrasts with a LAN or a VPN where the remote user would need to have Primavera installed on his or her laptop. Thin client computing can be very useful in allowing remote users access to computer programs. When a program is updated, all users in the organization have immediate access to the new program, and computer costs can be reduced by reducing the requirement for very high-powered laptops in the field. With the increasing use of wireless access, thin client computing can be very useful for construction managers.

■ Wireless Mesh Networks

Wireless mesh networks are a type of computer network that is beginning to emerge as a way of providing wireless access at the construction site. A Wi-Fi network has some drawbacks because it is subject to interference from engines of construction machinery. Additionally, as a facility is constructed and blocks signals from a hotspot, it may require additional hotspots or reconfiguration of existing hotspots. Additionally, each hotspot needs to have a hard-wired cable brought to it. A mesh network has characteristics that may solve some of these issues.

A mesh network provides a way to route broadband data wirelessly between nodes. Nodes are the electronic devices that send and receive data in a mesh network. A mesh network essentially replaces the need to wire the construction site with any data cables. For wireless access by computers in the field, a Wi-Fi access point can be attached to a mesh node or computers can be attached directly to the node device. A drawback to mesh networks is that there is currently no standard communications protocol, so you must use the same manufacturer's equipment for all of the nodes and required interface devices in a network.

The mesh network allows for continuous connections and reconfiguration around blocked paths by hopping from node to node until a communications path can be established. Mesh networks are self-healing, which is extremely desirable for construction applications. The network can still operate even when a node breaks down or a connection is lost (Wikipedia 2006).

In a wireless mesh network, nodes act as repeaters to transfer data to other nearby nodes. This allows a wireless mesh network to span large distances. Capacity can be added to the network by adding additional nodes. To provide a mesh network with Internet access requires only one node to have a hard-wired Internet connection, although additional nodes can be hard-wired with Internet gateways to provide redundancy. Mesh networks provide broadband connections, making voice and wireless video connections possible.

Figure 11-3 shows a typical wireless mesh network configuration for a construction project. Internet access is provided through an Internet gateway. Various mesh node devices are installed at strategic locations around the construction site to provide coverage. Users with laptops or PDAs can move around the construction site and get Internet access from Wi-Fi hotspots that are attached to mesh nodes. Other computers in fixed locations, like the field office, can be hardwired to nodes. Other devices like surveillance cameras are directly attached to nodes and send data over the mesh network.

Clearly, mesh networks are an emerging area of IT that can potentially make it easier to provide wireless Internet access to construction managers in the field. There are many manufacturers of mesh networking equipment.

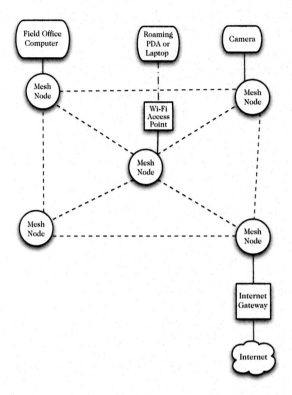

FIGURE 11-3 A Mesh Network Configuration for a Construction Project

■ WiMax: An Emerging Technology for Wireless Wide Area Networks

WiMax is an emerging broadband wireless technology. WiMax stands for "World Interoperability for Microwave Access," and IEEE 802.16 is the WiMax equipment standard. WiMax can work in both line-of-site and non-line-of-site situations, although the range of a WiMax system is much longer in line-of-site situations. In line-of-site mode transmission distances of up to 50 km are possible. For non-line-of-site applications, distances of 15 km around the base station have been observed. Data transmission rates can be up to 72 Mbps (Cayla et al. 2005).

The primary application of WiMax in the construction industry is to establish a wireless communication link between the company head office and remote construction sites. In many construction situations, it is difficult to provide a hard-wired link, such as a DSL or T1 line, and the WiMax network solves this problem. It is possible to deploy a WiMax network rapidly (Motorola says 24 hours for its Canopy system) which allows for the quick provision of data, voice, and video communications to the construction site. The WiMax network can be connected to a LAN in the field office and local Wi-Fi hotspots can be set up on the LAN allowing personnel on-site to communicate with the home office via the WiMax network (Westech Communications, Inc. 2005). Figure 11-4 shows a potential WiMax configuration for a construction company. The figure shows how the WiMax network can be used to connect the company's main office and its LAN and its Internet connection to the field office LAN and the Wi-Fi network in the field.

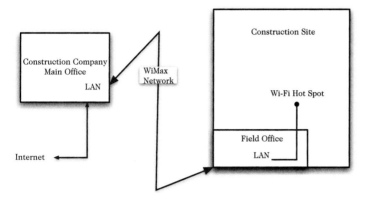

FIGURE 11-4 Using WiMAX to Connect Main Office with Construction Projects

■ Mobile Construction IT Applications

In this section we will look at a sample of the applications that are available for use in the field. We will discuss applications for data collection, extensions to well-known software packages for CAD and scheduling, and we will discuss possible knowledge management applications for mobile devices.

Data Collection Applications

Handheld computers have been successfully used in construction for field data collection. Some of the programs have been written for the activities of a construction contractor, whereas other field office software is available for use by government agencies, engineers, or architects supervising the construction of a project. Software aimed at construction contractors for use in field offices can organize project paperwork including daily logs, inspection reports, time cards, and cost control data. The software for project managers and owner organizations focuses more on inspector reports and collection of pay item data. A major component of field office management programs is the capability to allow data collection in the field using a PDA or laptop computer and using forms that are easy to fill out in the field. The handheld computers are taken to the work force to collect data, and communicate to the field office computer wirelessly, or they are brought back to the field office and downloaded to the office computer by a cradle and cable.

Appia FieldManager for Inspector Reports

An example of field data collection software is Appia FieldManager. FieldManager is software written to aid state and local municipalities in managing the field activities on road construction projects. Linked to FieldManager are FieldBook and FieldPad. This is software for laptops and handhelds, respectively, that allows for data collection in the field. Figure 11-5 shows how the screen of an inspector's daily report looks in FieldPad. In a stand-alone installation of FieldManager, the software and project database runs on a personal computer. FieldManager holds the data for all aspects of the construction project and accepts inspector's daily reports from the field. FieldManager can be used to initiate change orders and generate contractor's payments. FieldManager contains over 60 templates to generate the various documents and reports necessary for a highway project. Figure 11-6 shows the screen in FieldManager where pay items are posted from an inspector's daily report (Infotech).

FIGURE 11-5   FieldPad Inspector's Daily Report
Courtesy of: Field Manager® Software FieldBook® and Field Pad® screens are property of the Michigan Department of Transportation and Info Tech, Inc.

Trns•port FieldManager: A More Powerful Version of FieldManager for Large Projects

Trns•port FieldManager was developed as a partnership between Infotech (the developer's of Appia FieldManager) and the Michigan Department of Transportation. Trns•port FieldManager is a client-server application and is intended to be employed on large projects normally conducted by a state department of transportation. Trns•port FieldManager consists of a suite of programs for managing and tracking construction projects, documenting progress, initiating contractor payments, and communicating with an agency's central office. Trns•port Field-Manager is one of a suite of programs that are marketed by the American Association of State Highway and Transportation Officials (AASHTO) for use by transportation agencies. Details about all of the AASHTO software can be found at aashtoware.org.

The program can be used by state highway departments, engineering firms performing construction inspection, other agencies, and contractors. Appia FieldManager, which was discussed in the previous section, is intended for use on smaller projects run by smaller owner organizations like the engineering department of a small city. Like Appia FieldManager, Trns•port FieldManager includes FieldBook software for inspectors using

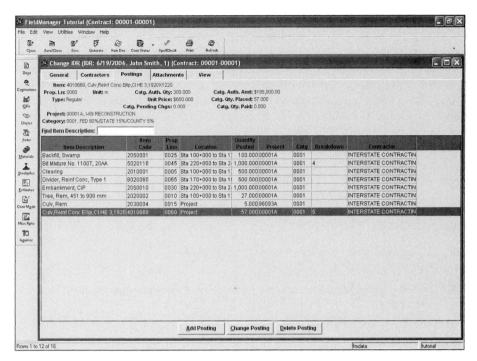

FIGURE 11-6 FieldManager Screen with Data Uploaded from the Field

Courtesy of: Field Manager® Software FieldBook® and Field Pad® screens are property of the Michigan Department of Transportation and Info Tech, Inc.

laptops and FieldPad software that works with PDA devices to collect project data. The Trns•port FieldBook program allows various items to be recorded at the project site including work items installed, contractor documentation, explanatory sketches by the inspector, and field notes. Many different state departments of transportation and engineering consultants are now using Trns•port FieldManager. Considerable cost savings have been documented by various users of this software. A case study documenting the successful implementation of Trns•port FieldManager by the Michigan Department of Transportation and the cost and person-hour savings follows at the end of this chapter.

Quality Management Using a Blackberry

A company called ATSG has developed a data collection system for residential constructors to monitor construction quality. The system employs the Blackberry wireless handheld device, or a tablet PC, and the Construction Quality Manager software. The Blackberry device is popular because of its small size, its ability to handle e-mails, and its full keyboard that makes data entry easier than some other types of handheld devices. This software is used by construction supervisors while walking through a house under construction. The Construction Supervisor records defects using the Blackberry device, and the recorded information is instantly sent to the central Construction Quality Manager database. Figure 11-7 shows a data input screen on a Blackberry device. Subcontractors are automatically notified of defects and the required corrective actions via e-mail or fax. The Construction Quality Manager software is either installed on a company's network or is hosted by ATSG (Advanced Technology Support Group). Greyson Homes, a homebuilder in the Baltimore/Washington area, uses the Construction Quality Manager software on its projects for quality assurance. It provides a database and management reporting system that provides Grayson Homes with a comprehensive view of its business. The company also reports that field supervisors find the built-in capabilities of the Blackberry (e-mails, calendar, and address book) to be extremely useful when used along with the quality assurance software (Blackberry 2005).

FIGURE 11-7 Inputting Construction Defect in the Construction Quality Manager System Using a Blackberry

Courtesy of: Advanced Technology Support Group Inc.©

Simple Punch List Software

The punch list software available for use in the field varies greatly in its complexity. A good example of an inexpensive solution for smaller contractors is the Punch List software for managing work tasks and punch list items that has been developed by Strata Systems (www.punchlist.com). This software provides a form to allow work tasks and punch list items to be recorded in the field using Palm devices. Software is also installed on a PC that links with the PDA to develop a database of work task items. With this software, a user can track list items, as well as send faxes and e-mails to project team members. Punch List can be used to collect work task and punch list information on projects and then faxed to project subcontractors by synchronizing with a PC in the office at the end of the day. A single license that allows the software to run on one PC and one handheld device costs around $300 (Strata Systems).

A Customized Application for Pile Driving

Ward et al. (2004) describe the use of wireless thin client devices to collect data about pile driving at two projects in Great Britain. Portable, battery-powered wireless access points were used at the site. Access points were mounted on cranes and equipment. Inspectors were given Pocket PC tablet devices to collect data and customized software was written to collect the data on a site-based server. Reasons for implementing the system were very interesting and apply to many construction firms:

- The need to reduce defective work due to incorrect information
- Errors translating data between the site and office
- A desire to promote the re-use of the collected data throughout the company

The results of the data collection effort were successful. It was estimated that the field data collection system reduced the requirement for remedial work by 75% on one project, a cost saving of £62,000 (approximately $115,000).

A Mobile Application to Provide Graphical Design Data for Highway Projects

Bentley has developed a software program for handheld computers that allows the project plan sheets to be viewed in the field and used for inspection and stakeout on highway and infrastructure projects. The program is called Inspector/Stakeout. The program's main features are:

- It provides easy access to project design features without the need to search through many files. Details of the design can be called up for reference. Figure 11-8 shows how the software displays the project plans and shows the design details for a culvert when it is selected by an inspector.
- When the mobile computer is also connected to a GPS device it can be used for project stakeout, eliminating the need for hand calculations of elevations in the field.
- For inspection, it allows notes to be referenced to the graphic design.
- It can allow as-built drawings to be created from field measurements using a GPS.

The Inspector/Stakeout program also provides written project documentation for use in the field including standard specifications, special provisions, standard drawings, and standard inspection procedures. The written documentation is linked directly to pay items shown on the graphical project plans. This type of program has tremendous potential to increase inspector productivity and reduce construction problems because it provides a construction inspector with all required project documentation on a handheld computer in an easily accessible format.

Mobile Versions of Popular Software

Most of the widely used software in the construction industry now offer mobile versions of software that run on PDAs. Increasingly, wireless access is possible using the mobile version of this software.

OnSite Enterprise

AutoDesk sells a program called OnSite Enterprise that provides digital plans and maps from a central server to Pocket PC devices. A program called OnSite View allows a PDA device to operate as both a client when working with the OnSite Enterprise server and as a stand-alone device. When using the PDA as a stand-alone device, OnSite View plans can be loaded onto the PDA by synchronizing with a personal computer. This powerful application allows computerized plans and GIS files to be viewed at the point of work. These programs allow AutoCAD drawing format (DWG) and drawing interchange format files (DXF) to be viewed and marked using a Pocket PC device. In addition, the OnSite software supports various GIS file formats.

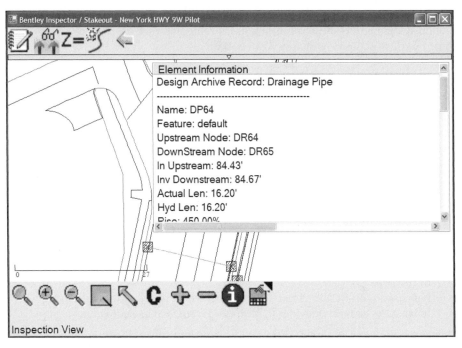

FIGURE 11-8 Showing Design Details Using Bentley Inspector/Stakeout
Courtesy of: Bentley Systems and Minnesota DOT

The program uses the pen-based interface of the Pocket PC to allow users to pan and zoom in real time to any view of the drawing. If the drawing contains layers, specified views that are saved within the drawing can be selected. In addition, the OnSite View program allows users to annotate plans by making free-hand drawings, using predefined symbols, and adding text. The markup files that are created on the Pocket PC can be moved back to a desktop PC and opened over a DWG or DXF file. Programs like this can allow easy reference to construction plans in the field. The ability to take notes and to modify the drawings can allow rapid generation of as-built drawings. It can also provide a way of notating construction problems in the field and marking their spatial location within the project (AutoDesk).

Mobile Scheduling

Mobile devices can be used in several ways to assist in scheduling. First, they can provide a way for personnel in the field to update activity information. Second, they allow managers at remote locations to access information about the status of projects. A good example is the mobile application that links to Primavera. The Primavera software has a handheld version called Mobile Manager, which runs on both the Pocket PC and the Palm devices. The software allows managers to both view and modify scheduling data in the field. Mobile Manager has several useful features. These include:

- The ability to transfer data from a handheld device to a computer either wirelessly or by connecting physically to a computer running Primavera.
- Users can select key items from a schedule for download to the handheld device.

- The capability to download data from multiple projects to the mobile device.
- Activities can be accessed and updated from a single screen. Figure 11-9 shows a handheld device displaying the data for an activity. Here a user could modify the activity in the field based on observed conditions.

FIGURE 11-9 Primavera Mobile Manager Activity Screen
Courtesy of: Primavera Systems, Inc.

Use of mobile devices for scheduling has several benefits. They include:

- Data no longer has to be entered twice. It goes directly from the handheld device into the company's scheduling system.
- Revised schedule data is more accurate because input comes from on-site observers and is recorded in real time.
- It can provide a way for managers outside the firm's office to access schedule details. Managers no longer have to return to the field office to determine the schedule status of construction operations.

Using Web-Based Applications in the Field

In Chapter 6, the capabilities of construction web portals were discussed. One of the main advantages of a web portal is that it is web-based and does not require any software on the user's computer. With the advent of wireless connections to the Internet, it is possible for a manager in the field with a properly equipped laptop computer to access all of the information contained in a construction web portal. Increasingly, as the use of wireless computing spreads in the construction industry, web portals will be accessed in the field to send and receive project data and information.

Many of the other computer applications we have discussed can be used in the field if web access is available. These include networked versions of scheduling and accounting software, content management systems, weblogs, and peer-to-peer applications such as Groove. It can be anticipated that the increasing speed of wireless data communications, as well as increasingly powerful and affordable computers, will fuel the increasing use of all types of web-based construction software at the construction site.

■ Collaboration and Knowledge Management at the Construction Site Using E-Books and the Internet

All of the applications we have discussed so far in this chapter have focused on the exchange of data and information. A growing issue of concern among construction researchers is ways to provide improved construction knowledge, and to promote collaboration between workers at the construction site. We must consider ways of providing mobile users information about how construction is accomplished to improve productivity and quality on projects. Knowledge can be provided by providing web access in the field and allowing field personnel to interact with weblogs and content management systems. Another possibility for providing managers and field supervisors with knowledge about the best ways to perform construction tasks is through the use of electronic books.

Electronic Books

Electronic books can be employed as a way of providing specifications, documents, and manuals of company practices. Several programs are available that allow for creation of these e-books for use on handheld devices. The primary software applications that are used to read electronic books are:

- Microsoft Reader. Microsoft Reader (Microsoft 2002) is free software distributed by Microsoft. It runs on laptops and PDAs using the Pocket PC operating system. The software for Microsoft Reader uses "ClearType Display Technology" that allows text to be clearly displayed. Microsoft Reader books can be created in several ways. One way is the Read in Microsoft Reader add-in for Microsoft Windows. This free software installs an icon in the Microsoft Windows toolbar that creates a Microsoft Reader file of the Word document when it is selected. Another way is to use the Reader Works Overdrive software to create the document. This software gives more flexibility in the creation of a book cover and formatting of the electronic book.

- Adobe Reader (Adobe). Adobe Reader comes in two different versions: one for Palm devices and the other for Pocket PC devices. The Adobe Reader software allows PDF documents created in Adobe Acrobat to be displayed on the handheld devices. Therefore, the documents created by Adobe Acrobat can be used on various types of mobile computers without modification. The Adobe Reader software works particularly well with tagged PDF files. Tagged PDF files can be "reflowed" to fit on the screen of a PDA (Johnson 2004). Tagged PDF files can be automatically created from text, such as a word processing file, but cannot be created from scanned images.

The electronic book allows personnel in the field rapid access to information, where in the past they may have needed to return to the project office to consult documentation that was too bulky to carry in the field. An alternative to electronic books is to provide web access and documentation in a content management system. However, the e-book is useful for projects for which no Internet access can be provided.

An example of a table of contents of a Microsoft Reader book is shown in Figure 11-10. The figure shows a screen capture from the laptop version of the program. An electronic book was developed that showed standard maintenance procedures required by the New Jersey Department of Transportation. This electronic book would be useful for construction contractors who perform maintenance work for the DOT. Foreman could use the e-book if questions arose about the correct procedures. Figure 11-11 shows a page from the maintenance procedure electronic book providing procedures for installing temporary steel plates on bridge deck patches.

FIGURE 11-10 Table of Contents for Maintenance Procedure eBook Using Microsoft Reader

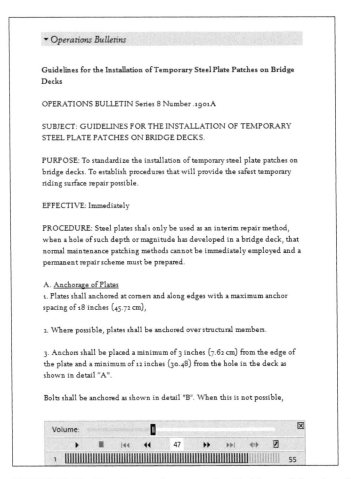

FIGURE 11-11   Repair procedures described in Microsoft Reader eBook

Using Web-Based Software to Promote Learning

In several chapters of this book, we have discussed how web-based applications such as Content Management Systems, weblogs, wiki, and peer-to-peer programs can promote collaboration between construction managers. Because wireless Internet access is rapidly becoming available at many construction sites, it is appropriate to suggest various ways these IT applications can assist managers in the field:

- Content Management Systems can be used to maintain documents explaining the construction company's standard operating practices and to provide educational information for inexperienced managers about the best practices for conducting various types of construction operations.
- Tools like weblogs and wiki can promote collaboration by allowing commenting on posts. Users in the field can post questions and comments and receive feedback from other members of the project team.
- Knowledge and information can be accessed without the need to return to the field office to look at manuals and documents.

Case Study: Implementation of Field Manager by the Michigan Department of Transportation

The Implementation of the Trns•port FieldManager system in Michigan provides a compelling justification for the application of mobile computers in the field. The benefits that have been obtained show mobile computers can streamline the flow of information on construction projects.

The Michigan DOT implemented the FieldManager system in part to cope with an increase in budget from $500 million to $1.5 billion while reducing staff from 5,000 to 3,000. The increase in efficiency that FieldManager provides allows inspectors that previously could work on only one project a construction season to inspect multiple projects. For very large projects, the number of inspectors on-site can also be greatly reduced using FieldManager (Overby 2002).

The implementation of Field Manager has:

...far exceeded initial expectation in terms of benefits provided to MDOT and to other state DOTs, local transportation agencies, and engineering-consulting firms who inspect or manage transportation construction projects. According to MDOT's most recent benefit analysis, FieldManager saves Michigan over $21.8 million per year (Computerworld Honors 2005).

Besides monetary savings, the time savings using FieldManager have also been substantial. The use of FieldPad to record inspection information rather than keying it into a laptop is estimated by the state of Michigan to save 13,300 hours annually of MDOT inspector time.

FieldManager has been able to generate the various time and cost savings because it has produced the following efficiencies (Computerworld Honors Program 2005):

- Recordkeeping is more accurate.
- Data collection is paperless. Inspectors no longer need to carry paper forms in the field, and the difficulty in reading handwritten forms has been eliminated.
- Instantaneous status reports can be produced, and payment request formwork can be processed rapidly.
- There is easier access to project information, both on- and off-site.
- Decisions can be made more rapidly.

In addition to the monetary and person-hour savings related to the use of FieldManager, the improved management information provided by the FieldManager program has allowed several major projects to be delivered ahead of schedule. "The M-6 a new 20-mile 'beltline' being constructed south of Grand Rapids in western Michigan, will be completed 3 years early... (Overby 2002)."

The success of the implementation of the FieldManager system using the mobile data collection programs FieldBook and FieldPad demonstrates how mobile computing, and the reduction of paperwork, can provide large cost savings to organizations willing to innovate.

Conclusions

Traditional construction software applications, such as scheduling and CAD, are extending their reach into mobile and wireless computing. Additionally, the capability to provide wireless Internet access at the construction site allows a wide variety of web-based software tools to be employed in the field. In this chapter we have explored how mobile computers can be used at the construction site to collect data, share information, update schedules, and view project plans. Ways to use mobile and wireless computing for knowledge management have also been suggested. The evolving wireless and mobile technologies are beneficial to the construction industry because they improve productivity by:

- Increasing the amount of time supervisors spend in the field and reducing trips to the field office to review documentation.
- Allowing updated information to be instantly shared with other managers and members of the project team.
- Reducing paperwork and data entry requirements for field supervisors.

Mobile and wireless technologies are developing rapidly. They have many beneficial uses in the construction industry and should be seriously considered for deployment by all construction companies.

Web Links to Equipment and Software

The following table is a listing of the web links for the various computer programs and equipment discussed in this chapter.

Application	Web Link
Thin Client Networking	
Citrix	http://www.citrix.com
Mesh Networks	
Firetied	http://www.firetied.com
Kiyon	http://www.kiyon.com/
Strix Systems	http://www.strixsystems.com/
Mobile Applications	
Appia FieldManager	http://www.infotechfl.com/software_solutions/fieldmanager.php
Trns•port FieldManager	http://aashtoware.org
Construction Quality Manager	http://www.atsgi.com/
Punch List	http://www.punchlist.com
Inspector/Stakeout	http://www.bentley.com/
OnSite Enterprise	http://usa.autodesk.com

Primavera Mobile Manager	http://www.primavera.com
Microsoft Reader	http://www.microsoft.com/reader/
Adobe Reader	http://www.adobe.com

■ References

Adobe. Adobe Reader for Mobile Devices. http://www.adobe.com/products/acrobat/acrrmobiledevices.html (Accessed November 27, 2005).

Advanced Technologies Support Group Inc. Construction Quality Manager. http://www.atsgi.com/PDF/CQM.pdf (Accessed November 17, 2005).

AutoDesk. AutoDesk OnSite Enterprise Overview. http://usa.autodesk.com/adsk/servlet/index?siteID=123112&id=700957 (Accessed November 27, 2005).

Blackberry. 2005. Grayson Homes Implements BlackBerry Wireless Communications System In Its Nationally Awarded Quality Assurance Program. http://www.blackberry.com/news/partner/2005/pr-12_01_2005-03.shtml (Accessed November 22, 2005).

Brown, K. 2005. Building Interest in the Construction Trade. http://www.wirelessweek.com/article/CA631033.html?spacedesc=Departments (Accessed November 18, 2005).

Cayla, G., Cohen, S., and Didier, G. 2005. WiMAX: An Efficient Tool to Bridge the Digital Divide, WiMAX Forum, http://www.wimaxforum.org/news/downloads/WiMAX_to_Bridge_the_Digitaldivide.pdf (Accessed March 2, 2006).

Computer World Honors. 2005. Field Manager for Construction Projects. http://www.cwheroes.org/Search/his_4a_detail.asp?id=4470 (Accessed February 26, 2006).

Dodd, A. 2005. *The Essential Guide to Telecommunications.* Upper Saddle River, NJ: Pearson Education, Inc.

Greengard, S. 2004. Building Knowledge Into Power. *iQ Magazine* V: 34–41.

Infotech. Appia Field Manager. http://www.infotechfl.com/software_solutions/fieldmanager.php (Accessed November 23, 2005).

Intel. Savings of 20 to 30 percent with Intel Centrino mobile technology help Laing O'Rourke build business. *Intel Business Center Case Study.* http://www.intel.com/business/casestudies/laing_orourke.pdf (Accessed November 16, 2005).

Johnson, D. 2004. What is Tagged PDF? http://www.planetpdf.com/enterprise/article.asp?ContentID=6067 (Accessed November 19, 2005).

Microsoft. 2002. Microsoft Reader-Download Source Materials and Conversions. http://www.microsoft.com/reader/developers/downloads/source.asp (Accessed November 28, 2005).

Overby, S. 2002. Paving Over Paperwork. *CIO Magazine.* http://www.cio.com/archive/020102/dot.html (accessed February 26, 2006).

Sandsmark, F. 2004. What You Need to Know about Wireless Networking. *iQ Magazine* 5 66–71.

Sinclair, J. and Merkow, M. 2000. *Thin Clients Clearly Explained.* San Diego, CA: Academic Press.

Strata Systems. Punch List-Countless Details-One Solution. http://punchlist.com/features.htm (Accessed November 24, 2005).

Ward, M., Thorpe, T., Price, A., and Wren, C. 2004. Implementation and Control of Wireless Data Collection on Construction Sites. *Electronic Journal of Information Technology in Construction,* 9: 297–311.

Westech Communications, Inc. 2005. Can WiMAX Address Your Applications? WiMAX Forum, http://www.wimaxforum.org/news/downloads/Can_WiMAX_Address_Your_Applications_final.pdf (Accessed March 3, 2006).

Wikipedia. 2006. Mesh Networks. http://en.wikipedia.org/wiki/Mesh_network (Accessed March 8, 2006).

Williams, T., Bernold, L., and Lu, H. 2006. A Survey of the Use of Wireless and Web-Based Technologies in Construction. In *Proceedings of the Tenth International Conference on Engineering, Construction, and Operations in Challenging Environments,* ed. Ramesh B. Malla, Wieslaw K. Binienda and Arup K. Maji, ASCE, CD-ROM.

200

CHAPTER **12**

Automation and Robotics in the Construction Industry

INTRODUCTION

Earlier chapters in this book have focused on how computer software can be used as a management tool to schedule projects, prepare bids, and allow construction managers to better collaborate. This chapter will focus on how information technology can be used to automate construction equipment and processes. It appears that the area of construction that has so far benefited the most from automation is heavy construction. The type of work done in heavy construction, such as placing and moving bulk materials, can be more easily automated than some of the labor-intensive tasks found in building construction that will require advanced robots to implement. In this chapter some of the current commercial applications of automation of heavy construction will be discussed. Then, some of the ways the construction of buildings can be automated using robotics will be discussed. Other emerging technologies like sensors, Radio Frequency Identification (RFID), and automated equipment monitoring will also be discussed in this chapter.

What Is Automation?

There are many different definitions of automation. Here are several definitions that provide a good sense of the meaning of automation in the construction industry.

- The art of making processes or machines self-acting or self-moving—Also pertains to the technique of making a device, machine, process, or procedure more fully automatic (STMicroelectronics).
- Making a process automatic, which eliminates the need for human intervention (IC Knowledge LLC).
- Substitution of human labor and skill with machinery, self-regulating devices, or computers (Pieter Nagel Logistics).

There are several levels in which automation can be implemented. At the first level, equipment can be added to construction machinery to provide improved feedback to the equipment operator. At the second level, computer equipment allows the construction equipment to operate automatically with little or no operator intervention. Finally, robotic equipment can be employed that is completely autonomous.

What Is a Robot?

In this chapter the potential uses of robots in construction will also be discussed. Like automation, robots have no universally agreed definition. However, it has been suggested that a robot must have several essential characteristics (MacDonald 2005).

- The robot must possess some form of mobility.
- The robot can be programmed to perform a variety of different tasks.
- The robot can operate automatically after programming.

Robots have already been used in the construction industry and their applications and potential will be discussed later in the chapter.

Reasons for Automating Construction

There are many motivating factors that drive the more widespread adoption of construction automation. These include (Khosnevis 2005):

- Competition—Automation offers the possibility of gaining an advantage over less automated competitors.
- Labor efficiency is low—Shortages of skilled labor-automation can reduce the need for labor in markets where there are labor shortages.
- Decrease hazards to human workers—Automated equipment can work in hazardous environments without exposing humans to unsafe situations.
- It is difficult to manage and control the construction site.
- Technological Advances—Like many of the technologies discussed in this book, new advances now make automation a possibility for many different types of construction operations that would be impossible to contemplate only a few years ago.
- To improve construction work quality

Benefits of Applying Automation

There are many areas in which automation can be beneficial. When construction processes can be automated there is the potential to:

- Reduce operation cycle times and increase productivity—Many construction processes can be done more rapidly when they are automated.
- Increase quality—Use of machines can improve the quality and uniformity of the finished product.
- Increase safety—Allow hazardous work to be automated to remove dangers to humans.

Barriers to the Implementation of Automation in the Construction Industry

Although there are many possible benefits to using automation in construction, there are some barriers that have so far blocked widespread adoption. These include (Khosnevis 2005):

- Unsuitability of some automated techniques for large-scale projects.
- The much smaller number of output units that are produced than in other industries.
- The use of conventional design approaches and materials that are not easily handled by automated equipment.
- Automated equipment is expensive.

■ Heavy Construction Applications

Various applications are now commercially available that allow heavy construction machinery to operate automatically with a significantly reduced need for operator intervention. It appears that the commercial adoption of automation for heavy construction has proceeded at a faster pace than for building construction. This has happened because it is significantly easier to automate a process, such as earthmoving, that involves moving bulk materials than building construction that involves a construction worker using various tools and techniques to install fixtures in a building. Applications include grading, excavation, and paving. In this section, products available from several manufacturers will be discussed.

Caterpillar Accugrade

The Accugrade GPS Control System is a machine control and guidance system that allows a dozer operator to grade with increased accuracy without the use of survey stakes (Caterpillar). GPS stands for Global Positioning System and is a worldwide radio-navigation system developed by the U.S. Department of Defense. GPS is satellite-based and provides suitably equipped users with three-dimensional position, velocity, and time information. The information transmitted by the satellites is available at no cost to users.

The Accugrade System uses the GPS data to deliver precise blade control and positioning data to the operator via a graphical display in the cab of the grading dozer. The equipment required to implement Accugrade GPS are a GPS base station that is located within radio range of the worksite, two GPS receivers mounted on masts on the grading equipment, and a radio receiver. A tilting sensor is also installed on the dozer blade. The in-cab display accepts flash cards that contain the design CAD documents that establish the required elevations. The Accugrade system is available for several tracked type Caterpillar tractors equipped with dozer blades.

Caterpillar also offers an alternative to using GPS for earthmoving. The Accugrade Laser Control System can be used for fine grading operations and operates using an off-board tripod mounted laser. The laser rotates and sends a thin beam of light that creates a grade reference over the work area. A laser receiver mounted on an adjustable mast on the grading equipment detects the laser signal. As the blade moves above or below finish grade correction information is sent to the in-cab display.

BOMAG Applications for Soil and Asphalt Compaction

BOMAG has developed applications for compaction that link GPS to computer-controlled compaction equipment (BOMAG Worldwide). BOMAG has produced applications for both soil compaction and asphalt compaction. BOMAG has developed equipment that, when installed on their vibratory rollers, can measure the compaction of both soil and asphalt.

For soil compaction, BOMAG produces a line of VARIOCONTROL rollers. The equipment senses areas with low or high bearing capacity and adjusts the amplitude of the vibrating roller accordingly. A device called the Terrameter provides a direct test for soil stiffness during construction using the relationship between the soil contact force and the deflection of the roller drum. The Terrameter indicates when further compaction is not possible and displays soft spots and non-uniform areas. A compaction measurement called the Evib (MN/m2) is calculated. Figure 12-1 shows the operation of the automated roller.

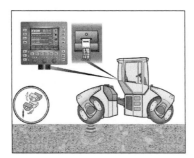

FIGURE 12-1   Automated Soil Compaction
Courtesy of: Bomag Americas, Inc.

BOMAG has also developed a compaction management system that links the compaction measurement equipment to a GPS receiver. Figure 12-2 shows a graphic of how the linkage between the automated equipment and GPS works. The system allows compaction measurements to be graphically visualized, displayed to the equipment operator, and stored for further analysis.

Figure 12-3 shows a drawing of the various components of the compaction management system including a laptop PC that can be used for data storage and analysis, operator displays, and printouts. The link to GPS allows exact pinpointing of locations where compaction levels may not be acceptable.

FIGURE 12-2 Automated Soil Compaction System Using GPS
Courtesy of: Bomag Americas, Inc.

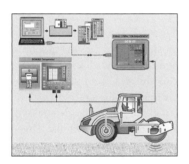

FIGURE 12-3 Compaction Management System
Courtesy of: Bomag Americas, Inc.

For the compaction of asphalt pavements using a tandem vibratory roller, BOMAG also provides automation equipment called Asphalt-Manager that varies the compactive effort of the vibratory rollers based on real-time measurement of the level of asphalt compaction. This can also be combined with GPS equipment to automatically provide a map of the area to be compacted to the operator showing the achieved level of compaction. Figure 12-4 shows a picture of the elements of the BOMAG asphalt compaction system.

Topcon Applications

Topcon manufactures devices that can be attached to construction equipment to provide a variety of functions. These include automated machine control for grading, excavating, and paving. They offer a selection of devices for grading and paving using laser, sonic, and GPS technologies. The equipment they manufacture includes:

- 3D GPS systems for machine control to various levels of accuracy depending on the application.
- 3D laser machine control.

Automation and Robotics in the Construction Industry

FIGURE 12-4 Automation of Asphalt Compaction Using GPS
Courtesy of: Bomag Americas, Inc.

- Automatic machine control for motor graders, dozers, excavators, pavers, and milling machines.
- Simpler indicate systems that inform the operator when he/she is deviating from the correct grade.

For GPS Control of grading equipment with low error tolerances, a GPS base station is set over a stationary point so it can accurately calculate its position. The moving machine also has a GPS receiver that is able to determine an approximate position. To grade to tighter tolerances, the base station sends signal corrections to the machine via a radio. These corrections help the machine determine its exact location on the site. At the machine, the data is processed by Topcon's 3D-GPS software to accurately provide the System Five 3D control box the real-time position of the machine in three dimensions. This positioning data is updated 10 times per second and, with the engineers' digital site plans, controls the blade of the machine for elevation and slope automatically (Topcon).

Topcon provides equipment that works with excavators and is useful for subsurface work. Using the automated excavator controls with GPS allows excavating equipment to be located precisely over utility centerlines. Use of the system allows the operator to view the location of excavator teeth in relation to the required grade in deep cuts, when working around structures and in blind excavations.

Paving is an area in which automation has been used for many years. Using various techniques, asphalt paving machines have long had the capability to work automatically by adjusting the paving machine controls to handle irregularities in the surface being paved and to produce a smooth, regular paving mat. This technology is continuing to advance, and Topcon now provides equipment that uses sonic technology to establish the correct grade and cross section of the pavement.

■ Automated Monitoring of Construction Equipment Location and Performance

The emergence of GPS and GIS technology is enabling construction companies to track and manage their equipment assets in real time. Typically, GPS transmitters are attached to construction equipment that sends location and vehicle performance data through wireless cellular or satellite data connections. System users can access equipment locations using a web browser that shows the location of the equipment on maps, drawings, or aerial photographs. The tracking systems are typically able to provide information beyond the location including the vehicles speed, ignition on/off, and dump up or down. The systems can also be used to deter theft by sending messages when the equipment leaves a certain geographical area that has been pre-defined by the system user. This application is known as "geo-fencing."

There are many applications and benefits of this technology:

- Improved cycle times for equipment, such as trucks that move off the construction site.
- The information captured by the software allows on-site inefficiencies and bottlenecks to be identified.
- Provides improved management data to maximize equipment use.
- Improved safety by providing notification when vehicles are speeding and notification when an equipment operator enters a hazardous area.

There are several providers of the software and GPS devices. They include the Trimble Construction Manager software (www.trimble.com/con_tcm.shtml) and the Qualcomm GlobalTracs equipment manager system (www.qualcomm.com).

Details of the Trimble Construction Manager Software

The Trimble Construction Manager software offers several interesting features for the location and management of mobile construction equipment, portable equipment (like generators and compressors), and personnel. They include:

- The Trimble Construction Manager software is a hosted service. Both the data that is collected in the field and the maps are hosted. The data from the field is hosted by Trimble, and the maps are hosted separately by a data partner, ESRI. Therefore the system does not require users to obtain any IT infrastructure or update map data. Users can access information from the system on a PC with an Internet connection because nothing is stored locally by default.
- Only one device is required to be attached to the vehicle. It sends the GPS data and has the capability to be wired to the machine to sense voltage changes in input lines to determine vehicle settings like on or off.
- The ability to modify settings such as speed limits, and to set up mileage and hours for maintenance settings. Figure 12-5 provides an example of an asset being added to the system, and an icon being designated for it.
- Figure 12-6 shows how a polygons can be created on a site map and equipment entrance and exits monitored.

- The location of assets can be viewed in real time on various types of maps, including color imagery, topographic maps, and streets. Figure 12-7 shows a color photograph of a site with icons showing the location of equipment and manpower.
- The Trimble Construction Manager software uses cellular data links to communicate with equipment. Mobile equipment sends data once per minute and portable equipment, such as compressors or generators can be set to send data once or twice a day.
- Various reports concerning equipment use can be generated by the Trimble Construction Manager software. They include an event and positions report, and a mileage and hours report. Examples of these reports are shown in Figures 12-8 and 12-9, respectively.
- The software works with Nextel Java-enabled cell phones to display the location of equipment to a manager using the phone in the field and to notify an employee when he/she is entering a hazardous area.

This example of the Trimble Construction Manager software shows how monitoring the location of construction equipment can be automated using the Internet, GPS, and GIS technology.

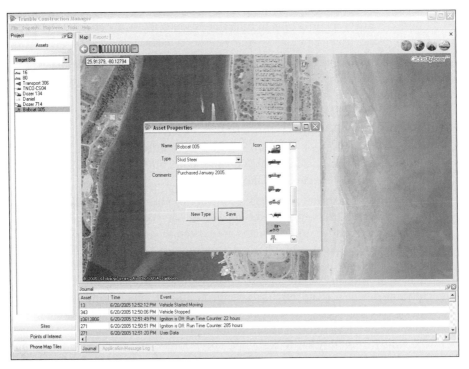

FIGURE 12-5 Adding an Equipment Icon Using Trimble Construction Manager
Courtesy of: Trimble

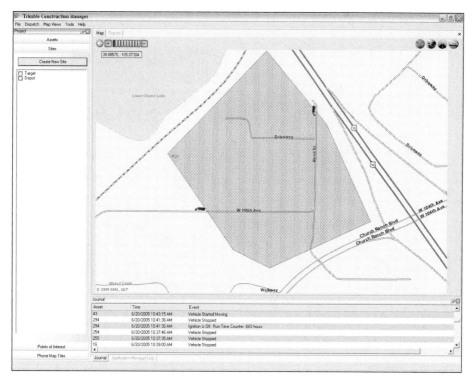

FIGURE 12-6   A Drawing with Polygons Showing a Geo Fence
Courtesy of: Trimble

FIGURE 12-7 A Photograph of a Construction Site Showing Equipment Location
Courtesy of: Trimble

EVENT AND POSITION REPORT

Report Period: 6/13/2005 12:00:00 AM to 6/13/2005 11:59:59 PM   *Mountain Standard Time*

Asset: Dump Truck 412

Site Visits	Speeding Events	Stop Events	Ignition Events
1	0	1	2

6/13/2005 - Asset Dump Truck 412

Time	Event	Odometer	Address
		miles	
5:43:33 AM	Ign On (Runtime 948 hrs) (Onsite Truck Sh	19938.8	8909 CENTER RD, SMALLTOWN, MA, 24706
5:44:56 AM	Position (Offsite)	19938.8	8909 CENTER RD, SMALLTOWN, MA, 24706
5:50:45 AM	Started Moving (Onsite Truck Shop)	19938.8	8909 CENTER RD, SMALLTOWN, MA, 24706
5:50:55 AM	Position (Offsite)	19942.0	4025 CENTER RD, SMALLTOWN, MA, 24706
5:51:43 AM	Leave Truck Shop	19942.6	4025 CENTER RD, SMALLTOWN, MA, 24706
5:51:55 AM	Position (Offsite)	19944.6	125 6TH ST, SMALLTOWN, MA, 24706
6:08:55 AM	Position (Offsite)	19947.6	601 7TH ST, SMALLTOWN, MA, 24706
6:09:19 AM	Enter Truck Shop	19947.6	8909 CENTER RD, SMALLTOWN, MA, 24706
6:09:55 AM	Position (Offsite)	19948.1	8909 CENTER RD, SMALLTOWN, MA, 24706
6:10:28 AM	Stopped Moving (Onsite Truck Shop)	19948.6	8909 CENTER RD, SMALLTOWN, MA, 24706
6:10:37 AM	Ign Off (Runtime 948 hrs) (Onsite Truck Sh	19948.6	8909 CENTER RD, SMALLTOWN, MA, 24706
6:10:55 AM	Position (Offsite)	19948.6	8909 CENTER RD, SMALLTOWN, MA, 24706

FIGURE 12-8 An Event and Position Report
Courtesy of: Trimble

MILEAGE AND HOURS REPORT

Report Period: 6/13/2005 12:00:00 AM to 6/20/2005 11:59:59 PM

Mountain Standard Time

Asset	Distance Driven	End Odometer	Run Duration	End Runtime Counter
	miles	*miles*	*hr:min:sec*	*hrs*
Fuel Truck 123	697.7	18172.6	73:15:19	532
Fuel Truck 456	701.7	20482.3	50:06:44	1034
Dump Truck 12	498.2	16170.6	30:24:17	991
Dump Truck 15	498.8	20500.3	51:14:29	1024
Dump Truck 15	501.6	26020.3	35:12:56	1362
Dump Truck 15	605.7	36020.8	26:21:14	1845
Dump Truck 15	128.0	18789.0	4:06:15	872
Dump Truck 15	326.6	21987.3	9:45:56	902
Dump Truck 15	458.3	11212.6	12:02:56	825
Pickup 21	188.6	69872.1	5:52:13	2035
Pickup 41	534.6	18274.5	61:02:50	1274
Water Truck 1	257.1	16888.7	21:37:27	825
Water Truck 2	643.4	14839.6	40:41:32	961
Lube Truck 49	174.5	9711.6	52:40:51	1225
Lowboy 87	801.9	23245.7	48:28:04	1524
Lowboy 89	423.6	15025.3	26:45:25	989
Lowboy 72	319.0	9658.5	23:48:50	765
Total	**7759.3**		**573:27:18**	

FIGURE 12-9 A Mileage and Hours Report
Courtesy of: Trimble

■ Robotics in Construction

Researchers in construction have experimented with the use of robots in the construction industry and there have been some commercial applications of robots. However, the high cost of robots and their complexity has been a barrier to their implementation in construction. Construction is fundamentally different from manufacturing where a robot can be installed on an assembly line and then work for years without requiring movement. However, in construction, projects can be relatively short-lived and the robot must be moved to the project and also moved within the project (such as between floors on a high-rise building).

Warszawski and Navon (1998) noted that conventional work in building construction is adapted to manual work, where a construction worker uses various materials and tools to complete a task. The complex motions of a human (finding, picking, placing, and attaching), and use of multiple tools are difficult for a robot to perform. They suggest that consideration must be given during design to use simplified building materials and systems that a robot can construct. Care must also be given to design a structure in which robots can move easily about the exterior and interior of a structure.

In the last 15 years several prototype robots have been developed. These earlier robotic implementations can be grouped into several areas:

- Materials handling—Robots employed in lifting large loads.
- Finishing robots—Robots employed in tasks, such as spray painting and concrete finishing, both in the exterior and interior of a building.

Japanese construction companies have been active in implementing robotic technologies. Japanese companies employed robots due to labor shortages, the desire to remain competitive versus international competition and the desire to market their use of advanced technology to potential clients. Additionally, large Japanese contractors have shown more of a willingness to invest funds for research and development than their American counterparts (Everett and Saito 1996). An example of the automation and robotic building construction techniques that have been employed in Japan is the SMART system (Kangari and Miyatake 1997). The system was employed to automate the construction of high-rise buildings. This technology included automated transportation systems, automated welding, placing of floor slabs, and an integrated information-management system. Extensive use was made of pre-fabricated components and simplified joints to facilitate the use of robots and automation.

With the practical application of robotic techniques to construction in Japan in the 1990s there was considerable interest in robotics by the U.S. construction research community, however, many prototype robotic systems have been developed but very few commercial applications have evolved. The economic downturn that occured in Japan in the late 1990s has also served to cool the environment for research into robotics in building construction. However, research continues, and someday with decreasing costs and increasing computer power robots will become more commonplace. Therefore, it is suggested that construction managers should monitor developments in this area.

Some of the recent work in robotics includes:

- Work at the National Institute of Standards and Technology has focused on the development of a test bed for researching robotic structural steel placement (Lyttle et al. 2004). NIST has experimented with a robotic crane fitted with a laser-based 3D site measurement system. The system is capable of autonomous path planning and navigation. Figure 12-10 shows autonomous steel docking using the robotic crane. More details about this research are available on NIST's Building and Fire Research Laboratory web page (http://www. bfrl.nist.gov/).

- Recent research has been performed concerning the possibilities for transferring robotic techniques from manufacturing to the construction industry. Koshnevis (2004) has discussed the potential to use contour crafting, an automated layered fabrication technology, to construct whole houses and their subcomponents in a single run.

FIGURE 12-10 NIST Robotic Crane for Structural Steel Erection

■ Tag and Sensor Devices

Developments in electronics now enable very small devices to be attached to materials to track their location and status. In addition, it is now possible to place sensing devices around the construction site and on construction equipment. Unlike robotics, these technologies are much easier to implement, and their use is likely to become widespread in the near future.

RFID Tags

RFID stands for radio frequency identification. RFID tags are small electronic tags to mark construction materials and are employed in the construction industry for tracking, sorting, and identifying construction materials. RFID tags are already finding widespread use for supply chain management in organizations such as Wal-Mart and the U.S. Department of Defense (Wyld 2005).

An RFID system consists of three components:

- Tags
- Readers
- Software to process and interpret the received data

The RFID tag serves as a unique identifier for the material it is attached to. The tag reader sends out a radio signal, and the tag responds with a radio signal identifying itself. The reader converts the received radio signal to data that is passed to a computer system capable of categorizing, analyzing, and acting on the identifying information. The ID tags consist of a chip, an antenna, and packaging. The ID tags are typically small and come in a variety of types depending on the application.

Several areas for the use of RFID tags have been identified. They include:

- Inventory control
- Material tracking and expediting
- Quality control and inspection

There are several beneficial impacts that the use of RFID tags can have in the construction industry. These benefits are:

- The ability to make better planning decisions. Important materials that might delay the critical path of a project can be tracked from fabrication to delivery at the project site.
- Better tracking of materials will improve productivity by ensuring that workers in the field receive necessary materials in a timely manner (Wood and Alvarez 2005).

Sensors

Electronic sensors have various applications in the construction and engineering. In particular, sensors have been employed within structures to measure structural performance. Various types of wired and wireless sensors are now available. Often these installations are aimed at measuring the long-term performance and deterioration of a structure. However, some sensors are available that relate directly to the construction portion of a facilities life cycle. Sensors have found their initial applications in the measurement of construction material quality.

IntelliRock II Sensors for Rapid Measurement of Concrete Strength

intelliRock II (www.intellirock.com) is an example of a commercially available sensor system to assist in concrete construction. The intelliRock system is designed to measure the temperature and maturity of concrete. The use of this type of sensor system can replace traditional strength estimation techniques involving destructive compressive tests on concrete cylinders. The traditional method of strength testing often required contractors to wait seven or more days to verify strength. The sensor system enables determination of the concrete strength in real time using the maturity method. The sensor system shows contractors the strength of the concrete in hours rather than waiting for days. The sensor system is designed to provide results that are unalterable and uninterruptible. intelliRock provides several benefits to construction contractors including:

- Improved construction workflow
- Earlier form removal
- Optimization of post-tensioning
- Improved quality control through the rapid identification of defective concrete
- Documentation of early strength and temperature data

The intelliRock II system is shown in Figure 12-11. The intelliRock II system consists of three components:

- The logger—The logger is a small sensor that is embedded in the concrete and is about the size of a 35-mm film canister. It contains a temperature sensor, a microprocessor, memory, a real-time clock, and a battery.
- A handheld reader—A handheld reader is used to gather temperature and maturity data from the logger by connecting to wires that lead from the embedded logger. The reader may be used simultaneously with a large number of sensors.
- Software—Software running on a PC accepts the collected data from the reader and provides analysis of the data.

The time savings possible from the use of a sensor system to more rapidly measure concrete strength illustrates how the application of sensor technology can impact construction projects by providing information more rapidly. A case study that describes the application of the intelliRock system is provided in the following section.

FIGURE 12-11 The intelliRock System
Courtesy of: Enguis, LLC

Wireless Sensors

The use of sensor technology, linked with wireless networking technology (discussed in Chapter 11) is now emerging as a way of collecting data automatically at the construction site. It is now possible to use small and relatively inexpensive sensors that could be placed on construction equipment or in various locations around a construction site. Micro-Electrical-Mechanical Systems is a technology that allows tiny mechanical devices, sensors, actuators, and electronics on a common silicon substrate (Wood and Alvarez 2005).

Sensors are available that can perform different functions, including:

- Temperature and humidity detection—Sensors can be placed on heavy construction equipment to determine if it is overheating.
- Acceleration tracking—Sensors can measure unusual vibrations in equipment.
- Motion detection—Sensors can be used to measure motion on a work site for security purposes.

The emergence of wireless communications now makes it much easier to deploy a network of sensors, and it can be expected that new applications for sensors will arise.

Case Study: I-10 Hurricane Damage Reconstruction Using IntelliRock II

The following case study describes the application of the intelliRock system on a fast-track construction project and describes how this sensor system was used to reduce the duration of construction. This case study also provides additional information about the maturity method of concrete strength determination. This case study is provided courtesy of Engius, the developer of the intelliRock system.

Hurricane Katrina slammed into the Mississippi and Louisiana Gulf Coast on Monday, August 29, 2005, resulting in numerous bridges being damaged or destroyed. The I-10 eastbound bridge over the Pascagoula River and adjacent marsh was badly damaged by two runaway barges.

The barges destroyed six of the bridge spans, rendering the eastbound bridge impassable. Mississippi DOT reacted in a timely fashion and on Wednesday, September 7 they received four bids from various contractors. The project was awarded to TL Wallace of Columbia, Mississippi for a 31-day contract to repair the bridge with a bid of $5.4 million. The contract called for $100,000 per day in potential damages. The same metric was used as the basis for determining the early completion incentive.

Getting Started

On Thursday, September 8, Mr. Mike Walpert and Mr. Clay Broom, TL Wallace's project managers along with Mr. Mike Ellis, the superintendent, arrived on site. The contract time clock began on Saturday, September 10 at 12:00 p.m. The demolition and removal work on beams, deck, barrier rails, piles, and pile caps began immediately.

continued

This project was somewhat unique in that Wallace chose not to move construction trailers to the site. Thus there were no hard-wired communication lines, such as faxes or desktop computers. Wallace worked directly off the remaining bridge deck and relied on portable generators, pickups as offices, and cell phones for communications. All work was scheduled on a continuous 24-hour rotation. Subsequent to the contract being awarded, Engius VP Richard Sallee contacted MDOT and the contractor regarding the possibility of utilizing the intelliRock system to help optimize work-flow. MDOT had allowed this technology to be used in other parts of the state on paving projects; however they had not utilized the technology for bridges. Mike O'Brien, with MDOT, had been contemplating the application of maturity on bridges for some time. This same technology had been utilized by numerous other state DOTs for bridge construction work. The most notable were the I-40 OK DOT bridge at Webbers Falls in the summer of 2002 and the I-20 bridge west of Pecos, Texas in 2003.

Concrete Maturity was first described in Europe in 1949 and it became an ASTM method in 1987 [ASTM C 1074]. The maturity method is a non-destructive, in-situ method that measures the extent of hydration, and subsequently the strength, of concrete by analyzing the time and temperature profile of the concrete.

Wallace contracted with Gulf Concrete, a Division of MMC Materials, Inc. to provide the concrete. Mr. Bobby Dowdy, regional QA manager for Gulf Concrete, had previously utilized the intelliRock system on commercial projects in the Mobile, Alabama, area and therefore was familiar with the multifaceted benefits the system provided (e.g., critical path management, quality control, and quality assurance enhancements).

Construction time was a critical factor on this project, and due to the chlorides associated with the salt water environment, concrete durability was a serious concern as well. Therefore it was imperative that in-situ strength be determined accurately and immediately to allow for timely stripping of forms. It was also important that the concrete be protected from moisture loss for proper cure to ensure durability.

Wallace and Gabe Faggard, the on-site project engineer for MDOT, were in agreement that concrete stripping strength would be determined by utilizing the maturity method. When the in-place strength was attained, the forms would be stripped and curing compound would be applied immediately to allow the concrete to continue to cure.

The mix design specified by the contractor and developed by Gulf Concrete called for a minimum strength of 2500 PSI in 12 hours. On Monday, September 19, Mr. Dowdy and Mr. Sallee began the process of developing a corresponding calibration curve for this mix design in Gulf Concrete's Pascagoula concrete laboratory. This calibration curve correlated the strength (in PSI) and corresponding maturity (°C-H) for a particular mix. The calibration indicated that forms could be removed when intelliRock indicated a maturity of 206 °C-H (2000 PSI) and 2500 PSI would be reached when intelliRock indicated a maturity of 231 °C-H.

Construction

The first concrete, Pier cap #81, was poured late in the afternoon on Tuesday, September 20. Mr. Broom first checked the intelliRock in-situ sensor at 9 hours and it read 318 °CH which corresponded to a strength of 3100 PSI. "Because of the mass associated with the pile caps the placed concrete was developing strength faster than anticipated" says Broom.

continued

On Wednesday, September 21 two pier caps, #79 and #80, were poured. On Thursday #77 was poured, and late night Friday, September 23 pier cap #78 was poured. On average, the caps were reaching the 2500 PSI strength in 7 to 8 hours. On Thursday, September 29 the entire 6-span bridge deck pour was accomplished. Guardrails followed on Friday.

Opening

Under the supervision of MDOT, Wallace opened the I-10 east span to full traffic on Saturday, October 1 at 2:00 P.M. TL Wallace had completed a challenging project 10 days earlier than the allotted contract time and thus earned $1 million in incentive bonus remuneration.

Mr. Ellis says, "many hard working Wallace team members contributed to this high-profile and demanding work and the associated work schedule. It took everyone from the laborer to the home office folks to pull this off, but especially the on-site, round-the-clock crew."

Wallace, a respected regional bridge contractor established in 1975, had not previously undertaken a project of this nature. It therefore presented a real opportunity for the 30-year-old bridge contractor. It was also their first experience utilizing the intelliRock concrete maturity system. "Wallace was very pleased with the results and the contribution intelliRock provided in the construction of the I-10 east span bridge, and we will definitely use this award winning technology on future jobs. It changes one's perspective on the ability to schedule and process the work flow," Mr. Broom says. "This is certainly a milestone in the history of TL Wallace Construction," says Mr. Ellis. "However, it is the traveling public and the residents of Mississippi Gulf Coast who are the real beneficiaries of this timely work."

Project Timeline Summary
August 29th: Hurricane Katrina
September 7th: Bids Received
September 8th: Contractor Arrived On-Site
September 10th: 31 Day Clock Started
September 19th: intelliRock Calibration
September 20th: Concrete Pours Start
September 29th: Deck Complete
October 1st: Bridge Open (10 days early)

■ Web Cameras to Automatically Record Construction Progress

The Internet now enables users to remotely view pictures or streaming videos of construction sites. This technology has several benefits. They include:

- The ability to keep project managers abreast of construction activities and progress even if they are at locations geographically removed from the construction site.
- The ability to create a visual archive of project progress, activities, and important events.

- Improve communication between project participants by allowing the owner, designer, contractor, and the public to access pictures and videos of project activities.
- The systems are web-based and no special software is needed to view images. Anyone with a web browser and the proper permission can view pictures and videos.
- Web camera systems can be employed as both a collaboration tool for project participants and as a way to inform the public of project progress.
- Web cameras can be used as a marketing tool to show potential customers performance on past projects.

Typically a contractor must obtain a camera and provide electrical and high-speed Internet connections. The cameras that are employed are controllable by users viewing the pictures on the web and can tilt, zoom, and pan to view areas of the site that are of interest. There are several companies that provide fee-based hosting of a web site that provides a user interface to the output from the camera. These companies are popular because a contractor can view and use the pictures and video without having to host any software on his/her own servers. Several services are available that provide construction web camera services. They include TrueLook (www.truelook.com), OxBlue (oxblue.com), and OnSite-View (www.onsiteview.com).

The TrueLook Web Camera System

In this section, we will explore the capabilities of one of the web camera services, TrueLook, in depth. When using TrueLook a contractor must purchase a telerobotic camera (or cameras) that can tilt and pan so it is compatible with TrueLook's system. All other aspects of the system are hosted and maintained by TrueLook.

Figure 12-12 shows The TrueLook user interface (Truelook). The interface includes various features including:

- View, tilt, and pan controls. A user is able to move the camera and zoom in on areas of interest.
- Multiple users can access the system simultaneously.
- Views from multiple cameras
- Still, streaming, and time-lapse view modes
- The ability to annotate, archive, and e-mail pictures. TrueLook has a collaboration tool that allows users to capture a specific image and e-mail it to others. Recipients of the e-mail can click on the picture and link to the live view.
- Images can be maintained in a web-accessible searchable database to provide a visual archive of project progress.

Figure 12-13 shows the control panel for the TrueLook system. Through this interface, it is possible to:

- Control access to the pictures and video
- Enable and disable features
- Schedule archive shots to create time-lapse videos
- Build image catalogs
- Create panoramic views
- Manage archives

This look at the details of TrueLook illustrates how the increasing speed of Internet connections is enabling new technologies that allow construction progress to be viewed from remote locations by project participants and the public.

FIGURE 12-12 The Truelook User Interface
Copyright: True Sentry, Inc.

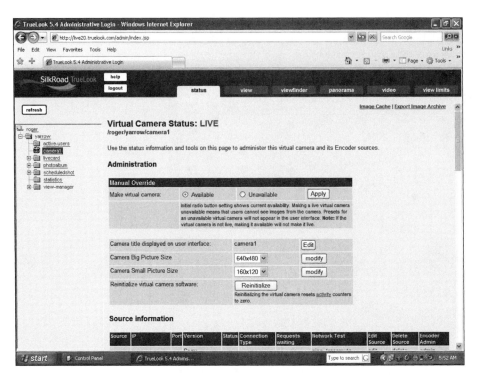

FIGURE 12-13 The Truelook Control Panel
Copyright: True Sentry, Inc.

■ Sources for More Information About Automation

New developments in automation occur frequently. Several web sites can provide the reader with updates about new developments and more information about construction automation. Of particular interest to constriction contractors is the FIATECH consortium. FIATECH is a non-profit consortium that funds research in various areas of automation and information technology in construction. Their web site, listed in Table 12-1 contains interesting information and documents about automation developments. NIST has already been discussed in this chapter and its web site describes various types of robotics and automation research in construction. The International Association for Robotics and Automation in Construction (IAARC) provides a web page with information and links about construction robotics. This organization publishes a directory of commercially available construction robots and automation equipment.

TABLE 12-1 Websites With More Information About Construction Automation

Organization	URL
FIATECH	http://www.fiatech.org/
NIST	http://www.bfrl.nist.gov/
IAARC	http://www.iaarc.org/

Conclusions

Although the pace of adoption of automation in the construction industry has been slow, the use of automated equipment is increasing. Automation of heavy construction machinery is now a reality. Due to the high costs of robots, and the complexity involved in robotizing many human tasks, automation has been slower in building construction. Research is continuing in the use of robots in construction and, as the technology evolves, robots will begin to be more widely employed. It is suggested that contactors monitor developments in this area to quickly learn of new commercial technologies.

Summary of Web Links

This table provides web links to the automation equipment, software, and sensor devices discussed in this chapter.

Caterpillar	http://www.cat.com
BOMAG	http://www.bomag.com
Topcon	http://www.topcon.com
Trimble	http://www.trimble.com/con_tcm.shtml
Qualcomm	http://www.qualcomm.com
intelliRock	http://www.intellirock.com

References

BOMAG Worldwide. Compaction Measurement and Documentation Systems. http://www.bomag.com/worldwide/index.aspx?fm=%2fconstruction_machineries%2fmeasuring_systems.aspx&DID=100000000&Lang=10000 (Accessed December 22, 2005).

Caterpillar. Accugrade. http://www.cat.com/cda/layout?m=37483&x=7 (Accessed December 23,2005).

Everett, J.G. and Saito, H. 1996. Construction Automation: Demands and Satisfiers in the United States and Japan. *Journal of Construction Engineering and Management* 122:147–151.

IC Knowledge LLC. Glossary of Integrated Circuit Terminology. www.icknowledge.com/glossary/a.html (Accessed December 27, 2005).

Kangari, R. and Miyatake, Y. 1997. Developing and Managing Innovative Construction Technologies in Japan. *Journal of Construction Engineering and Management* 123(1):72–78.

Khoshnevis, B. 2004. Automated Construction by Contour Crafting-Related Robotics and Information Technologies. *Automation in Construction* 13:5–19.

Lyttle, A., Kamal, M. Saidi, S., Bostleman, R.V., Stone, W.C., and Scott, N.A. 2004. Adapting a tele-operated device for autonomous control using three-dimensional positioning sensors: experiences with the NIST RoboCrane. *Automation in Construction* 13: 101–118.

MacDonald, Chris. What is a Robot? http://www.ethicsweb.ca/robots/whatisarobot.htm (Accessed December 28, 2005).

Pieter Nagel Logistics. Glossary. http://www.pnl.com.au/glossary/cid/28/t/glossary (Accessed December 27, 2005).

STMicroelectronics. Glossary. http://www.st.com/stonline/press/news/glossary/glossary.htm (Accessed December 26, 2005).

Topcon. TPS Machine Control and Equipment Automation. http://www.topconmc.com/index.html (Accessed December 29, 2005).

Truelook. Truelook~Construction Cams. Construction Web Cam. http://www.truelook.com/solutions/enterprise/benefits.htm (Accessed February 26, 2006).

Warszawski, A. and Navon, R. 1998. Implementation of Robotics in Building: Current Status and Future Prospects. *Journal of Construction Engineering and Management* 124(1): 31–41.

Wood, C. and Alvarez, M. 2005. Emerging Construction Technologies: A FIATECH Catalogue. http://www.fiatech.org/links.htm#vet (Accessed January 2, 2006).

Wyld, D. 2005. *RFID: The Right Frequency for Government.* Washington, DC: IBM Center for the Business of Government.

222

CHAPTER **13**

A Roadmap for Construction IT Implementation

INTRODUCTION

In this book, several techniques for using computers to manage construction projects and companies have been discussed. Depending on a construction company's situation, various paths are available to implement the use of information technology in construction. In this chapter we will consider the issues and suggest a possible approach to implementing IT solutions in construction companies. Certainly, any successful IT implementation in construction must be in alignment with a company's goals and must address the real needs of the construction company's management team and project managers in the field.

■ Defining the Goals of an IT Project

The goals of any IT project must be clearly defined. The following sections address some important issues that must be considered when defining the IT solution to be implemented.

Knowledge or Information?

One of the first questions that must be asked when considering implementing IT within your firm is to identify the main purpose of the system. It seems clear that the types of web-based systems that have been discussed in this book divide into IT systems for the management of information and systems for the management of knowledge. Some of the IT systems we have discussed include systems that can contain both information and knowledge. For example, a weblog can contain posts of knowledge about how to perform construction and posts of information and data about a project.

Clearly, information management has been at the forefront of IT development to date. As has been discussed, there are many web portal services available for the management of information on complex projects. However, it is becoming increasingly obvious that construction companies must learn to capture their internal knowledge for reuse on future projects. In the future, IT managers will need to identify both the information and knowledge management requirements of their firm, and will need to select appropriate information technology tools.

When seeking to implement a new computer system, it is appropriate to ask if the application is for knowledge or information exchange. In the construction context, the focus of information exchange is to manipulate documents related to the firm, and the individual construction projects that the firm conducts. For knowledge management systems, the focus is transferring knowledge about how a construction company conducts its operations.

For example, in a construction information system, a construction foreman may be provided with a mobile computer to fill in informational documents like daily logs and reports. A knowledge management system for the foreman may alternatively provide him with a mobile computer linked to a weblog or an e-book that contains the firms "lessons learned" about the best techniques for asphalt paving.

What Are Your IT Resources and Capabilities?

An important factor that dictates what IT solutions can be implemented is the construction firm's current use of computers, current capabilities for using computers, and ability to accept and adapt to changing technology. Construction companies exist in a broad spectrum of computer use. Organizations can be described as being in different states of informatization. Informatization is defined as an organization's ability to use and absorb information technology. Various metrics can be defined to measure the level of informatization within a firm. Example metrics include measurements of a firm's IT strategy, IT resources, IT use, and IT performance. To measure IT performance data such as the number of computers per employee can be collected. Use statistics can be gathered on hours of computer use per day. The performance of a firm in implementing IT can be measured by the total number of IT projects it has undertaken (Lim 2001).

Although computer usage has been increasing in the construction industry, many firms still use computers in only rudimentary ways. Therefore, it behooves the management of a construction company to consider where their firm is situated on the spectrum of informatization portrayed in Figure 1-3. If your company has a low level of informatization, then a complex IT system will be difficult to implement because your firm does not have enough experience in the use of IT.

Certainly there is also a relationship between the size of the construction company and the level of informatization that is achievable. Large firms conduct large projects and require complex IT systems to manage the projects effectively. Additionally, large construction companies often have the resources to develop specialized IT systems for project management. However, there are many solutions available to smaller firms, and applicable to smaller-scale projects.

Many of the software programs used in construction require training. Many small firms that have only a few employees cannot spare crucial employees to give them the required training in a new IT system. Therefore, a small firm should look to implement IT solutions that can be relatively easily learned and assimilated into the firm's operations.

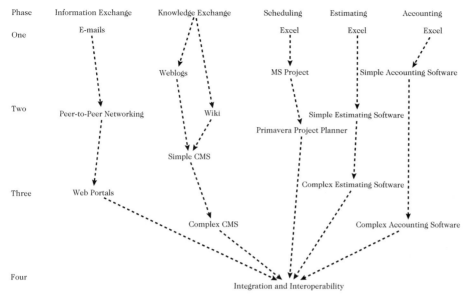

FIGURE 13-1 IT Implementation Guidelines

What Size Projects Do You Do?

The size of projects that a company performs clearly has a major influence on the types of IT systems that are appropriate. The many web portal service providers have arisen in response to the need to control the documentation on complex construction projects. However, for smaller projects many "general-purpose" solutions are available. There are tremendous possibilities to bring computers and web-based communications to areas of the construction industry, mainly small projects, where computers have never been used.

For small projects, it is appropriate to consider easy-to-implement solutions such as the peer-to-peer networking capabilities of software like Groove or a weblog for both information and knowledge exchange. There are also simple estimating and scheduling packages available that are applicable to smaller projects that are reasonably easy to implement and learn.

Considering IT Tools for Large Projects

Of course, the size of projects a firm undertakes is highly correlated with its size and capabilities. For large firms, the primary consideration in selecting software is its "scalability." That is, it must consider if the estimating software can handle thousands of cost codes. The scheduling software must be robust enough to handle thousands of activities for a project, and perhaps multiple projects simultaneously. Any web portal selected must be able to handle thousands of change orders and project correspondence. It must also be able to handle hundreds of daily users.

Partners

Several of the IT solutions discussed in this book, particularly construction web portals, allow other project partners to participate in the IT system. Consideration must be given to the computer capabilities of the other participants in the construction process. Personnel from the designer, owner, subcontractor, and suppliers, may all participate in an IT system. Rivard et al (2005) have discussed case studies of several construction projects for which the use of IT solutions such as web portals could not be extended to subcontractors because they lacked the computer expertise.

As a general contractor, you must consider if particular subcontractors have the computer expertise to participate in the web portal. Perhaps a subcontractor's computer expertise can even become a factor in the selection of a subcontractor. In Chapter 1 of this book we discussed how construction contractors exist in a wide range of levels of informatization. Various metrics can be defined to measure the level of informatization within a firm. Example metrics include measurements of a firm's IT strategy, IT resources, IT use, and IT performance. To measure IT, performance data such as the number of computers per employee can be collected. To measure use, statistics can be gathered on hours of computer use per day (Lim 2001). The total number of IT projects it has undertaken can measure the performance of a firm in implementing IT. Possibly a general contractor or owner can solicit these statistics as part of the selection process of project team members.

■ Implementation Phases

Companies exist in different phases of IT competence and experience. In this section these phases are described and suggestions are given for how a company can execute an IT strategy. Figure 13-1 shows a suggested roadmap for IT implementation. Construction companies travel through several phases as they implement new computer technologies. An incremental approach to the implementation of new technologies has been found to work best in construction. It is unlikely that a firm that has never used computers can easily assimilate the use of web portals or handheld devices in the field. Therefore, the IT roadmap shown in Figure 13-1 suggests four phases in which a construction firm exists. Depending on the construction company's IT phase, the following paragraphs suggest IT solutions that a firm may wish to consider.

Phase 1

The first step in bringing computer and IT technology into a construction firm is to move from paper-based documents to using the computer to generate documents. A starting place for many is to implement the Microsoft Office suite. Excel is a good general-purpose program that can be used to prepare estimates and display simple schedules. Possibly, Microsoft Access databases can be used to store historical cost information. Project documentation is generated using Microsoft Word.

The next step in the implementation of computers within the construction firm is to provide Internet access. Internet access provides the ability to send and receive e-mails and to access the World Wide Web E-mail allows text messages as well as documents to be attached and exchanged. E-mail provides the rudimentary beginnings of both knowledge and information management.

Phase 2

The second phase of computer implementation for the construction company should focus on beginning to use web-based software to provide information and knowledge exchange. This book has discussed various web-based IT solutions that require little technical knowledge other than the ability to use a web browser. A good starting point to increase collaboration and improve information and knowledge management is a weblog. In particular, a weblog that has the capability to include files as attachments to posts can provide a good way of sharing information between project team participants.

Basic accounting and scheduling software is available that is reasonably easy to learn and implement. Although Primavera Project Planner is considered the industry standard for scheduling, Microsoft Project and Primavera Suretrak can be used for smaller projects. For accounting and project cost control, simple programs such as QuickBooks and Peachtree software can be used by small firms.

Another option for firms using the Internet is to use peer-to-peer programs such as Groove to form small networks for project collaboration and information exchange. This type of peer-to-peer networking is attractive as a means of bringing sophisticated IT capabilities to a small construction company at low cost, and without the necessity of having a significant computer infrastructure beyond personal computers.

Phase 3

In Phase 3 a construction company has accumulated enough expertise within the firm to successfully consider implementing web portals for complex projects that require significant collaboration between other project participants, and as an aid in tracking project correspondence and documentation. At this phase, it is possible to consider more sophisticated accounting applications, and integration of accounting with estimating that a program such as Timberline Estimating can provide. Additionally, a firm that is becoming sophisticated in the use of computers could consider implementing mobile applications to collect field data or for displaying e-books of construction information and knowledge.

Phase 4

The final stage in the evolution of IT use in the construction company is to fully embrace the interoperability standards that are emerging. Primavera Project Planner and Expedition are interoperable, and a schedule can be created from the data input to the estimating program. Interoperability also allows data taken from CAD documents to be used to automatically generate paperless estimates.

Companies in Phase 4 can begin to consider the use of advanced techniques such as 4-D modeling to consider alternative ways of constructing facilities and to determine potential conflicts during construction. The potential is also increasing to provide wireless access to construction sites, both in urban areas and remote sites. A company in Phase 4 has the computer sophistication to work with service providers to implement new types of wireless networks that are suitable for use on construction sites for information and knowledge exchange.

Putting It All Together: A Case Study of Advanced IT Use In Construction

The following case study illustrates how a construction contractor with sophisticated IT capabilities can use computers in a variety of ways to improve productivity and reduce costs. The case study is composed of excerpts from a June 11, 2001 article in *ENR* magazine by Andrew G. Roe titled "Subcontractor Customizes IT Upgrades to Push Growth."

One Florida specialty contractor has seen firsthand the challenges in making information technology management a corporate priority. But executives say that the gain has been well worth the pain. MSI-Encompass, formerly Mechanical Systems Inc., Orlando, began a deliberate effort several years ago to boost its IT savvy. "About eight years ago, our company was basically computer illiterate," says Roger Scherer, a senior project manager. MSI founder and President William M. Dillard, who initiated the IT revamp, says improved technology is helping his company stay competitive. "Typically, our projects are much faster than they used to be," he says. The IT surge coincides with rapid growth for MSI, which specializes in heating, ventilation and air conditioning, plumbing, electrical systems, and building maintenance services. The company grew from $28 million to $38 million in revenue between 1997 and 1999. That year, it was also acquired by Houston-based Encompass Services Corp., a nearly $4 billion specialty contractor. Dillard says Encompass invests close to $25 million annually in its own IT management.

Project management and document control had previously been handled using disparate systems. MSI implemented Prolog from Meridian Project Systems companywide. It was supplemented by Primavera Project Planner and Microsoft Project for scheduling. Estimation Inc. software is used for cost estimating. Microsoft Office is used for word processing, spreadsheet and database applications, and Autodesk Inc.'s AutoCAD for shop drawings. "Standardization is key," says Turner. "There were too many different systems in place, and that doesn't lend itself to data sharing."

When off-the-shelf software doesn't work, MSI builds custom applications in-house. Using programming tools built into Microsoft Office products, Scherer has developed macros that help track rental equipment and employee assignments the way the company prefers. "It's usually faster for me to develop a custom database than to teach someone in the field to use another system," he says.

Using a combination of in-house and off-the shelf software, MSI reaped the benefits on its work at Orlando Sea World's 32-acre Discovery Cove project. "We had over 150,000 man-hours in 16 months," says Scherer. "We were never off by more than two to three people" in projections, out of a staff that sometimes reached 90 employees, about 25% of its work force. Like other contractors, MSI works on numerous jobsites simultaneously and needs to exchange information quickly and often. To accommodate that, it has been adding more portable computing devices in the field. About 25% of superintendents now use laptops, and service technicians are beginning to use personal digital assistants to manage contact and equipment information. PDAs are being used as stand-alone units and are not yet configured to communicate with the network or other MSI computers, says Turner.

But laptops and other field computers are being linked to the network to share programs and data. Using MetaFrame from Citrix Systems Inc., MSI sets up field computers to run selected programs hosted at the company's main office via the Internet. "It allows anyone with computer access to access a server-based application across the Internet or a wide-area network," says Turner.

◼ Conclusions

When considering the implementation of IT in the construction industry, it is necessary to consider a firm's capacity to accept the new technology. Construction companies can be considered to exist in four phases of IT use. For companies interested in using IT, it is suggested that they begin at Phase 1 to acquire computer capabilities and incrementally increase their use of web-based systems and advanced technologies.

◼ References

Lim, Soo Kyoung. 2001. A Framework to Evaluate the Informatization Level. In *Information Technology Evaluation Methods and Management*, ed. Wim Van Grembergen, 144–153. Hershey, PA: Idea Group Publishing.

GLOSSARY

This glossary defines information technology terminology and acronyms that are used in this book.

4D CAD   A three-dimensional visualization of a structure linked to the construction project schedule. It allows users to view a simulation of the construction process over time.

802.11   The IEEE standards to which WiFi wireless networking equipment is built.

5D CAD   A 4D CAD visualization that also includes cost and estimate information for project elements.

Automation   Making a process automatic by substituting human labor or skill with machinery, self-regulating devices, or computers.

BMP   The designation for Bit Map Files, which are a form of digital image file.

Broadband   Describes a high-capacity internet communications link that can handle large amounts of data, voice, and video.

Browser   A computer application that allows a user to view and interact with information on the World Wide Web.

CAD   CAD (computer-aided design) software is used by architects, engineers, and constructors to create precision drawings or technical illustrations. CAD software can be used to create two-dimensional (2-D) drawings or three-dimensional (3-D) models.

Cellular Network   The wireless voice and data service provided by telephone companies. It consists of a radio network divided into fixed cells, each with a fixed transmitter.

Client   A computer that is linked to a central server in a client-server computer network.

Client-Server Network   A computer network in which a central server computer handles requests for information and data from linked client computers.

Community of Practice   A community of practice can be defined as a learning network in which practitioners connect to solve problems, share ideas, set standards, and develop relationships.

Concept Map   A diagram illustrating knowledge in a particular domain through the links between various concepts. Concept mapping is an effective way of representing a person or group's understanding of a problem domain.

Construction Web Portals   A collaborative web site that allows project participants to generate and exchange construction project plans and documentation.

Content Management System   A web-based tool for organizing information and knowledge. A content management system has the capability to publish, modify, index, search, and retrieve documents and information.

Critical Path Method (CPM)   The critical path method is a network scheduling technique that is widely used in the construction industry. Most scheduling software employs the critical path method.

Document Management   The management of discrete documents and images throughout their life cycle. In construction this includes the management of design documents from the initiation of design through the construction and use of a project.

DSL   Digital Subscriber Line. It is a technology for providing high-speed data communication to small businesses. It shares the same phone line as telephone service but does not interfere with voice communications.

DGN   The standard file format of CAD documents created using Bentley Microstation and Intergraph MGE software.

DWG   The standard file format for CAD files created using AutoCAD.

DXF   An AutoCAD file format that is used to import and export data between CAD systems.

Electronic Bidding   Web-based systems that allow construction project bids to be submitted using the World Wide Web.

Electronic Book Multimedia programs that run on personal computers and PDAs and resemble the format of a traditional book. They are an effective way of providing manuals and other textual information to users at the construction site.

Firewall Computer hardware or software that protects a computer network from unauthorized access by users in other networks.

FTP FTP (File Transfer Protocol) is a common method of moving files between Internet sites.

Geo Fencing The process in which GPS can be used to determine when construction equipment has left or entered a designated geographic boundary.

GIS A GIS (geographic information system) enables you to envision the geographic aspects of a body of data. It lets you query or analyze a database and receive the results in the form of some kind of map. Since many kinds of data have important spatial aspects, a GIS can have many uses: weather forecasting, population forecasting, and land use planning, to name a few.

GPS The GPS (Global Positioning System) is a satellite-based system that makes it possible for people with ground receivers to pinpoint their geographic location.

Hosted Service Web-based computer programs that are hosted on another company's servers. The end user pays a fee to the hosting company for the service.

Hot Spot A wireless access point that sends and receives radio signals from mobile computing devices like laptops and PDAs.

HTTP The communications protocol that enables web browsing.

Hypertext Markup Language (HTML) The coding language used to create web pages for use on the World Wide Web.

Information Technology IT (information technology) is a term that encompasses all forms of technology (computers, software, and telecommunications) used to create, store, exchange, and use information in its various forms (business data, voice conversations, still images, motion pictures, and multimedia presentations.

Informatization The level of an organization's ability to assimilate and use Information Technology.

Internet Collection of public computer networks that are interconnected across the world by telecommunications links. Allows users to communicate with each other using software interfaces such as electronic mail, FTP, Telnet, and the World Wide Web.

Interoperability This refers to the ability of different computer programs to work with each other. In particular, the ability to embed cost and scheduling data in CAD documents is emerging to allow construction project information to be input in a single format.

JPEG A common file format for images.

Knowledge Management Knowledge management can be defined as the processes in which knowledge is created, acquired, communicated, shared, applied, and effectively utilized.

Line-of-Balance Schedule A location-based scheduling technique. It shows the location of crews over time and will show where crew conflicts occur.

On-line Bidding Web sites that can be used to obtain bidding information like project line items, and that also can be used to submit project bids.

On-line Plan Room A web site that can be accessed to download CAD files of project plans.

Mesh Network A type of wireless network that is capable of routing broadband data to network nodes. It can eliminate the need for cables to set up WiFi hotspots.

Monte Carlo Simulation A type of simulation where a large number of trial runs are performed, and the probability of possible outcomes are determined. It is used in construction to calculate the probability of a particular project duration.

Myql An open-source database program that must be installed with many open-source Content Management System programs.

Local Area Network (LAN) A group of computers and other computer devices such as printers and scanners that are connected within a limited geographic area.

Open Source Software Computer software that is often distributed for free and that allows users to modify the original source code.

PDA Personal Digital Assistants are handheld computing devices. The most common types are Pocket PC, Palm, and Blackberry devices.

PDF   The file format for CAD drawing and documents created using Adobe Acrobat.

PHP   A software program that allows web developers to create dynamic content for web pages. PHP is used to develop web-based software applications.

Peer-to-Peer Network   A type of computer network in which each workstation has equivalent capabilities and responsibilities. There is no central server.

Podcasting   Audio files that are delivered by an RSS feed.

Reverse Auctions   Auctions in which the low bid is visible to all bidders. Bidders may modify their bids until the auction closes.

Robotics   The science of using machines to perform manual functions without human intervention.

RSS   Really Simple Syndication. A format for notifying users of new content at a web site.

RFID   RFID (radio frequency identification) is a technology that uses radio signals to transmit data. An RFID system consists of three components: an antenna and transceiver (often combined into one reader) and a transponder (the tag). The antenna uses radio frequency waves to transmit a signal that activates the transponder. When activated, the tag transmits data back to the antenna.

Server   A computer that delivers information and software to other computers linked by a network.

T1   A standard line for high-speed data communications that connects businesses to their internet provider. T-1 circuits operate at 1.54 million bits per second.

Tablet Computer   A computer with a touch-sensitive screen that allows users to input data using a stylus and allows users to make and store hand-written information electronically.

TCP/IP   A collection of protocols that define the basic functions and operation of the Internet.

Thin Client   A computer that gets its computing resources from a server. A laptop user with a thin client connection can use and run a computer program that resides solely on the server.

TIFF   A standard computer file for images.

Virtual Private Network   An arrangement that allows connections between offices, remote workers, and the Internet without requiring private lines.

Web-based System   Computer software accessed by a user using a web browser. The user does not need to have the software installed on his/her computer. Typically web-based systems are hosted on a web server.

Weblogs   Weblogs are web-based computer systems that allow small chunks of information to be posted to a web page without knowledge of HTML or computer programming.

Web Portal   A web site or hosted service that offers a broad array of computing resources. In a construction context, web portals are popular for exchanging project documents and improving collaboration.

Wide Area Network (WAN)   A group of LANs that communicate with each other, possibly widely separated geographically.

Wiki   A web site that allows users to build content collectively. No knowledge of programming is required to add content or create additional linked pages.

Wireless network (WLAN)   A wireless LAN is one in which a mobile user can connect to a local area network through a wireless (radio) connection.

WiFi   Wireless Fidelity. A term for wireless networks that conform to the IEEE 802.11 family of standards.

WiMax   A wireless access technology providing greater coverage distances than WiFi technology.

World Wide Web (WWW)   Part of the Internet for which connections are established between computers containing hypertext and hypermedia materials.

XML   XML stands for Extensible Markup Language. It allows software designers to create customized tags that allow the definition, transmission, and interpretation of data between software applications.

INDEX